PUBLISHERS' NOTE

As its title implies, the Series in which this volume appears has two purposes. One is to encourage the publication of monographs on advanced or specialised topics in, or related to, the theory and applications of probability and statistics; such works may sometimes be more suited to the present form of publication because the topic may not have reached the stage where a comprehensive treatment is desirable. The second purpose is to make available to a wider public concise courses in the field of probability and statistics, which are sometimes based on unpublished lectures.

The Series was edited from its inception in 1957 by Professor Maurice G. Kendall, under whose editorship the first 21 volumes in the Series appeared. He was succeeded as editor in 1965 by Professor Alan Stuart.

The publishers will be interested in approaches from any authors who have work of importance suitable for the Series.

CHARLES GRIFFIN & CO. LTD.

GRIFFIN'S STATISTICAL MONOGRAPHS AND COURSES

No. 5, formerly *Characteristic Functions* by E. Lukacs, is now published
independently of the Series

For a list of other statistical and mathematical books see back cover.

STOCHASTIC POINT PROCESSES

AND THEIR APPLICATIONS

S. K. SRINIVASAN, M.Sc., Ph.D.

Professor of Applied Mathematics
Indian Institute of Technology
Madras

Monograph No. 34

Series Editor:

ALAN STUART

GRIFFIN LONDON

CHARLES GRIFFIN & COMPANY LIMITED
42 DRURY LANE, LONDON WC2B 5RX

Copyright © S. K. SRINIVASAN, 1974

All rights reserved

First published   1974

ISBN: 0 85264 223 7

Demy Octavo, xi + 174 pages
5 line illustrations

Set by E W C Wilkins Ltd London N12 0EH
Printed in Great Britain by Whitstable Litho Straker Brothers Ltd

Dedicated to

MY BROTHER

PREFACE

Point processes form a special class of discrete-valued stochastic processes and have acquired special importance mainly for two reasons. First, there is a huge variety of applications of point processes covering a wide spectrum ranging from classical physics and control systems to biology, cybernetics and management science. Second, it is in this limited class of stochastic processes that the probabilist can get a glimpse of the structural relations of non-Markov processes. The theory of point processes had its origin in the attempts by Yvon and Von Mises, who studied the properties of certain functionals associated with random points, though the current terminology of point processes is due to Wold who introduced in a formal way the stochastic process of point events. Point processes have been studied in great depth by many individuals in connection with specific problems arising from different natural phenomena such as random noise, cosmic ray showers, kinetic theory, population growth, and telephone traffic. However, these results should be viewed in the wider perspective of the theory of stochastic processes. Monographs on stochastic processes do not, in general, deal with point processes, except perhaps in the context of branching phenomena. Of course a lucid account of the renewal process and of some of its generalizations is given in *Renewal Theory* by Cox, *The Statistical Analysis of Series of Events* by Cox and Lewis, and in *Mathematical Methods in the Theory of Queueing* by Khintchine. The present monograph goes much farther in its scope, covering both general theory and applications. The author has preferred to resort to a heuristic account of the theory, mainly to make the contents of the book accessible to a wider range of users, although some parts of the book can be presented through rigorous analytical methods.

Chapter 1, which is introductory in nature, deals with the general scope of application of point processes in different fields. Introductory material leading to simple point processes is also presented in this chapter. This is followed by a general account of the theory of point processes. Chapter 3 deals with renewal processes and their ramifications which include alternating renewal processes and Markov renewal processes. The properties of stationary point processes are then presented in detail. The fifth chapter deals with doubly stochastic Poisson processes, while the sixth contains an account of multivariate point processes. The last three chapters deal with some of the applications of point processes to Statistical Physics, Management Science, and Biology. The choice of the material in these areas of application is severely limited by the author's experience and interests.

The reader is assumed to have a knowledge at the level of Feller's *An Introduction to Probability Theory and its Applications*, volume I. The material in each chapter is illustrated by examples, and problems have been added at the end of most of the chapters. It is hoped that the book will be of interest to graduate students specializing in Applied Probability and Statistics. Specialists working in theoretical areas like Statistical Physics, Biomathematics and Operational Research may also find the book useful, especially while attempting to make probabilistic models.

The author is grateful to M. Jaya Kumar for typing the manuscript and preparing it for the printer. Thanks are also due to Drs S. Kalpakam and A. Rangan for their help in the preparation of the manuscript and the correction of the proofs.

S.K. SRINIVASAN

Madras,
June 26, 1973

CONTENTS

1   INTRODUCTION

1.1   Role of the theory of point processes

The main charm of point processes lies in a huge variety of applications, ranging from classical physics and biology to modern control theory and cybernetics. Of late, applications of the results of point processes have been found to be increasingly useful in the realms of management science. It is interesting to note that during the International Statistical Year 1970–71, as many as six international conferences dealt with either the theory or the application of point processes to physical, or social, or biological phenomena. Before we embark on a formal (and yet heuristic!) account of the theory of point processes, we will first attempt to explain how a wide range of different phenomena in diverse fields such as statistical physics, population growth, communication and control theory, and management science have contributed to the growth of the subject. We will then give a very brief account of some of the important concepts used in the theory of stochastic processes.

STATISTICAL PHYSICS

Some of the important aspects of the theory of point processes were first developed in connection with the study of kinetic theory, response phenomena, and the theory of cosmic ray showers.

Kinetic theory of fluids   A drop of fluid can be considered as an assemblage of a large number of molecules distributed over a six-dimensional phase space of configuration and momentum. If $(\mathbf{p}_i, \mathbf{q}_i)$ represents the momentum and coordinate of the ith molecule, the momentum of the drop of fluid $\mathbf{P}$ can be written as

$$\mathbf{P} = \sum_i \mathbf{p}_i,$$

where the summation extends over the entire assembly constituting the drop. The pressure tensor $\mathcal{P}_{\alpha\beta}$ due to the drop of fluid when it is not too dense is given by

$$\mathcal{P}_{\alpha\beta} = \sum_{ij} (\mathbf{p}_i - \mathbf{P})_\alpha (\mathbf{p}_j - \mathbf{P})_\beta.$$

It has been experimentally observed that the velocities or moments of individual molecules are not deterministic and are subject to certain probability laws. The occurrence of the events corresponding to the molecule being found in an infinitesimal region of phase space surrounding the point $(\mathbf{p}, \mathbf{q})$ can be identified as a point event in the six-dimensional phase space. The

2

stability of the expected values of $\mathbf{p}$, $\mathcal{P}_{\alpha\beta}$, and other macroscopic quantities like temperature is one of the major objectives of a macroscopic theory. This problem will be highlighted in Chapter 7.

Response phenomena   The phenomenon of shot noise is, historically, one of the earliest examples of a response due to a series of point events (on the time axis) corresponding to short pulses, each pulse occurring with the passage of an electron from cathode to anode. The cumulative response due to the series of pulses is an observable quantity, and this can be identified as a linear functional(*) of the stochastic process corresponding to the point events.

Cosmic ray showers   In cosmic ray showers, the main problem of interest is to study the energy spectrum of particles hitting a particular layer of the atmosphere. Here again, in developing a theory, we have to deal with a statistical assemblage of particles distributed over energy, the energy distribution arising from a branching phenomenon exhibited by high-energy particles in their passage through the rare matter of the upper atmosphere. The product-density approach due to Bhabha (1950) and Ramakrishnan (1950) was essentially a by-product of the attempt to estimate the fluctuation of the number of particles about the mean number.

POPULATION GROWTH

The stochastic problem of population growth has continued to attract the attention of probability theorists during the past one hundred years. In 1874, Watson and Galton formulated the problem of extinction of families. In the forties of the present century the problem received further impetus, partly because of an increased interest in the theory of stochastic processes and its applications, and partly because of the analogy between population growth and other multiplicative processes occurring in the realms of nuclear reactions and of cosmic rays. In 1949, D.G. Kendall made a fundamental contribution to the theory of point processes in general and of age-dependent population growth in particular; the cumulant generating functional and its functional derivatives, introduced by him, have turned out to be extremely useful tools in the study of population point processes. Some of the biological phenomena for which the theory of point process provides useful models are (i) bacteriophage production, (ii) the extinction of bacterial colonies due to viral attack, (iii) the phenomena of carcinogenesis, and (iv) control of communicable diseases.

(*) A functional is a scalar-valued function defined on a linear space.

COMMUNICATION AND CONTROL THEORY

A large number of problems in communication and control theory arise as responses (linear as well as non-linear) to point events. Here essentially we have optimization problems, the quantities to be optimized being functionals of one or more point processes. A good deal of work has been done by using the techniques of time-series analysis. Especially in problems involving estimation, analysis by the point-process approach will prove to be extremely useful.

MANAGEMENT SCIENCE (OPERATIONAL RESEARCH)

The use of point process models in management science is of fairly recent origin. The fields in which point process models are particularly useful are queues, inventories, and reliability theory.

Queues Although the theory of queues is fairly old and well established, its connection with the theory of point processes is of comparatively recent origin and is mainly due to D.G. Kendall (1951, 1964) and Khintchine (1955). Feller (1966) focussed the attention of workers in this field on the need for a general unified technique based on the theory of renewal point processes.

Inventory theory The theory of inventories deals with storage problems in general, and during the last decade several models of inventories based on renewal theory have been proposed and studied. A systematic account of these models can be found in the monographs of Arrow, Karlin and Scarf (1958) and Hadley and Whitin (1963). Inventory theory has also been studied from the viewpoint of water storage problems. Pioneering efforts in this direction were made by Moran (1954, 1959). Here again the theory of renewal processes plays a significant role in obtaining the statistical characteristics of the process.

Reliability theory Reliability theory deals with general methods of evaluating the quality of systems subject to gradual deterioration or failure. It establishes the statistical frequency of occurrence of defects in devices, and also methods of prediction. The epochs of total failure and recommencement of normal operation of the machine due to completion of repair constitute point events. In fact, recent work on reliability theory suggests the need for enlargement of the theory of point processes of one type of events so as to include processes in which events of different types occur. This leads to an interesting area known as multivariate point processes (see, for example, Cox and Lewis (1970)).

1.2 Stochastic processes

A stochastic process is simply a collection of random variables X_t $(t \in T)$. The collection is an ordered collection if T is one-dimensional. Thus the stochastic process $\{X_t: t \in T\}$ is uniquely specified if the joint distributional properties of the collections $\{X_{t_1}: t_1 \in T\}$, $\{X_{t_1}, X_{t_2}: t_1, t_2 \in T\}$, ... are specified. Let us assume that T is one-dimensional. For practical purposes we can divide stochastic processes into two classes according to whether T is discrete or not. Notice that if S is the sample space of events, then

$$X_t: S \xrightarrow{X_t} A_t,$$

where A_t is a subset (not necessarily proper) of R, the reals[*]. This enables us to define the state space as

(1.2.1) $$\mathcal{S} = \bigcup_{t \in T} A_t.$$

We can subdivide the stochastic processes according as to whether $\mathcal{S}$ is discrete or not. Thus we have, broadly speaking, four categories:

   (i) T discrete, $\mathcal{S}$ discrete;
   (ii) T discrete, $\mathcal{S}$ continuous;
   (iii) T continuous, $\mathcal{S}$ discrete;
   (iv) T continuous, $\mathcal{S}$ continuous.

A stochastic process of type (i) is completely specified by the joint probability mass functions $\pi_n(i_1, i_2, \dots, i_n; j_1, j_2, \dots, j_n)$ defined by

(1.2.2) $$\pi_n(i_1, i_2, \dots, i_n; j_1, j_2, \dots, j_n)$$
$$= \Pr\{X(t_{j_1}) = x_{i_1}, X(t_{j_2}) = x_{i_2}, \dots, X(t_{j_n}) = x_{i_n}\}$$

for all $t_j \in T$, $x_i \in \mathcal{S}$, $n \geqslant 1$. On the other hand, a process of type (ii) is completely specified by the joint distribution functions $\rho_n(x_1, x_2, \dots, x_n; j_1, j_2, \dots, j_n)$, where

(1.2.3) $$\rho_n(x_1, x_2, \dots, x_n; j_1, j_2, \dots, j_n)$$
$$= \Pr\{X(t_{j_1}) \leqslant x_1, X(t_{j_2}) \leqslant x_2, \dots, X(t_{j_n}) \leqslant x_n\}$$

for all $t_j \in T$, $x_i \in \mathcal{S}$, $n \geqslant 1$, or equivalently by the joint density functions $\pi_n(x_1, x_2, \dots, x_n; j_1, j_2, \dots, j_n)$, where

(1.2.4) $$\pi_n(x_1, x_2, \dots, x_n; j_1, j_2, \dots, j_n)$$
$$= \lim_{\Delta_1, \Delta_2, \dots, \Delta_n \to 0} (\Pr\{x_1 < X(t_{j_1}) < x_1 + \Delta_1; x_2 < X(t_{j_2})$$
$$< x_2 + \Delta_2 \dots x_n < X(t_{j_n}) < x_n + \Delta_n\})/\Delta_1 \Delta_2 \dots \Delta_n.$$

Similar remarks apply to situations (iii) and (iv).

In what follows let us assume that $t_{j_1} < t_{j_2} \dots < t_{j_n}$. The law of multiplication of probabilities enables us to express the joint probability density (or mass) functions (1.2.4) (or 1.2.2) in terms of the conditional density (or

---

[*] It is easy to extend the idea when A_t is a subset of R^n.

mass) functions. For example, we have

(1.2.5)
$$\rho_n(x_1, x_2, \ldots, x_n; j_1, j_2, \ldots, j_n)$$
$$= \rho_1(x_1, j_1)\, \rho_2'(x_2, j_2 | x_1, j_1)\, \rho_3'(x_3, j_3 | x_2, j_2; x_1, j_1) \ldots$$
$$\rho_n'(x_n, j_n | x_1, j_1; x_2, j_2; \ldots; x_{n-1}, j_{n-1})$$

(1.2.6)
$$\pi_n(i_1, i_2, \ldots, i_n; j_1, j_2, \ldots, j_n)$$
$$= \pi_1(i_1; j_1)\, \pi_2'(i_2, j_2 | i_1, j_1)\, \pi_3'(i_3, j_3 | i_2, j_2; i_1, j_1) \ldots$$
$$\pi_n'(i_n, j_n | i_1, j_1; i_2, j_2; \ldots; i_{n-1}, j_{n-1}).$$

It should be noted that the conditional density functions are equally difficult to deal with. Great simplification is effected if a further restriction is imposed on these conditional density functions. Thus it is convenient to assume that

(1.2.7)
$$\rho_n'(x_n, j_n | x_1, j_1, x_2, j_2, \ldots, x_{n-1}, j_{n-1}) = \rho(x_n, j_n | x_{n-1}, j_{n-1})$$
$$(t_{j_1} < t_{j_2} < \ldots < t_{j_n}),$$

(1.2.8)
$$\pi_n'(i_n, j_n | i_1, j_1, i_2, j_2, \ldots, i_{n-1}, j_{n-1}) = \pi(i_n, j_n | i_{n-1}, j_{n-1}).$$

A stochastic process which possesses the above property is known as a *Markov process*. In such a case (1.2.5) reads

(1.2.9)
$$\rho_n(x_1, x_2, \ldots, x_n; j_1, j_2, \ldots, j_n)$$
$$= \rho_1(x_1, j_1)\, \rho(x_2, j_2 | x_1, j_1)\, \rho(x_3, j_3 | x_2, j_2) \ldots \rho(x_n, j_n | x_{n-1}, j_{n-1})$$

Thus a Markov stochastic process is completely specified in terms of the two functions $\rho_1(..)$, $\rho(..|..)$. Similar statements can be made about processes of types (i), (iii) and (iv). In case (iv) we have a relation identical with (1.2.9) except for the difference that t_j varies over a continuum. If $\rho(x, t | y, t')$ is the corresponding conditional density function, i.e.

(1.2.10)
$$\rho(x, t | y, t') = \lim_{\Delta \to 0} \Pr\{x < X(t) < x + \Delta | X(t') = y\}/\Delta,$$
$$(t > t'),$$

then it is easy to see that for any τ satisfying $t' < \tau < t$, we have by the application of the basic laws of probability,

(1.2.11)
$$\rho(x, t | y, t') = \int_x \rho(x, t | x', \tau)\, \rho(x', \tau | y, t')\, dx',$$

a relation named after Chapman and Kolmogorov. The arbitrary nature of τ has been extensively used to obtain various differential equations satisfied by $\rho(..|..)$ and corresponding to different limiting forms of $\rho(., t | ., t')$ as $t \to t'$. A good account of the differential equations and their solutions is given by Bharucha-Reid (1960) and Ramakrishnan (1958). Many interesting properties of $X(t)$, especially for processes of type (i) and (iii), are of great practical importance. For an account of such processes, known as *Markov chains*, the reader is referred to Kemeny and Snell (1960) and Parzen (1962).

Relationships of the type (1.2.7) or (1.2.8) can be expressed in the following manner for any general stochastic process $X(t)$. A stochastic process $X(t)$ is Markovian in nature if the probabilistic structure of $X(t + \Delta)$ for arbitrary Δ depends only on that of $X(t)$ and is independent of the probabilistic structure of $X(u)$ $(u < t)$. Non-Markovian processes arising from the residuary class of stochastic processes not satisfying the above requirement are indeed difficult to deal with directly. In many cases it is convenient to define a new stochastic process with additional elements which render the resulting process Markovian. Some simple examples were constructed and analysed by Cox (1955) (see also Srinivasan (1955)). From an intuitive physical point of view, it is reasonable to visualize a non-Markovian process as the projection of a Markovian process, the non-Markovian nature arising from the absence of the extra "coordinates" lost by projection.

There is another dichotomy in stochastic processes that is of great significance from the point of view of theory and applications. The probabilistic structure of $X(t)$ may be independent of the origin of reference of t. In other words, the probabilistic structure of the stochastic process may be invariant under arbitrary translation of t. In such a case the stochastic process is said to be *stationary*, the non-stationary processes forming the residuary class. There are several degrees of stationarity and we shall mention a few of the important ones, *confining our attention to case (iv)*. A process $X(t)$ is said to be *simply stationary* if $\pi(x, t)$ defined by

$$(1.2.12) \quad \pi(x, t) = \lim_{\Delta \to 0} \Pr \{x < X(t) < x + \Delta\}/\Delta$$

satisfies the condition

$$(1.2.13) \quad \pi(x, t + h) = \pi(x, t) \quad \text{for arbitrary } h \text{ for all } x \in \mathcal{S}, t \in T.$$

A stochastic process is *completely stationary* if the joint density function $\pi_m(x_1, x_2, \dots, x_m; t_1, t_2, \dots, t_m)$ defined by

$$(1.2.14) \quad \pi_m(x_1, x_2, \dots, x_m; t_1, t_2, \dots, t_m)$$
$$= \lim_{\substack{\Delta_i \to 0 \\ i=1,2,\dots,m}} \Pr \{x_i < X(t_i) < x_i + \Delta_i, \quad i = 1, 2, \dots, m\}/\Delta_1 \Delta_2 \dots \Delta_m$$

satisfies the condition

$$(1.2.15) \quad \pi_m(x_1, x_2, \dots, x_m; t_1 + h, t_2 + h, \dots, t_m + h)$$
$$= \pi_m(x_1, x_2, \dots, x_m; t_1, t_2, \dots, t_m)$$

for arbitrary h and for all $x_i \in \mathcal{S}, t_i \in T, i = 1, 2, \dots, m$ and $m = 1, 2, \dots$. Very often the parameter t signifies time, and in such a case the origin is usually chosen at $-\infty$. Sometimes we may not visualize the evolution of the process during the infinite past, and in such a case stationarity would mean invariance under arbitrary but restricted finite time translation. In practical situations it is convenient to impose (1.2.15) only for $m = 2$, in which case the process is said to be *stationary to second order*. For an account of the general properties of such processes, the reader is referred to Bartlett (1966) and Parzen (1962).

Special stochastic processes of type (iii) have very many interesting properties, especially if the sample paths $X^s(t)$ are non-decreasing functions of t with probability one. In such a case, by a relabelling of the indices that characterize the set $\mathcal{S}$, the random variable can be assumed to take only positive integral values. The random variable $X(t)$ then can be identified with processes in which random points are realized on a one-dimensional continuum, so that $X(t)$ represents the count of the random points up to t (starting from an appropriate origin). The properties of such processes were studied by Wold (1948) who called them *point processes*. The object of this monograph is to provide a systematic account of stochastic processes of this general type. Before we embark on a formal study, however, it will be useful to describe some simple examples. The first of these is the Poisson process, and this we consider in section 1.3. In the following section (1.4) we discuss the Yule–Furry process which provides an example of a branching process.

1.3 The Poisson process

There are several equivalent ways of defining a Poisson process. A convenient starting point is the imposition of stationarity on the discrete-valued stochastic process $X(t)$ of type (iii) defined in the previous section, the set $\mathcal{S}$ being identified with the set of positive integers including zero. Introducing the random variable

$$(1.3.1) \quad Y(t_1, t_2) = X(t_2) - X(t_1) \quad t_2 > t_1,$$

we note that complete stationarity implies that $\pi'(n, t_1, t_2)$ defined by

$$(1.3.2) \quad \pi'(n, t_1, t_2) = \Pr\{Y(t_1, t_2) = n\} \quad n = 0, 1, 2, \ldots$$

is a function of $t_2 - t_1$ only. Thus, writing

$$(1.3.3) \quad \pi(n, t_2 - t_1) = \pi'(n, t_1, t_2),$$

we observe that the Chapman–Kolmogorov relation (1.2.11) in this case takes the form, for $t > 0$,

$$(1.3.4) \quad \pi(n, t) = \sum_{m=0}^{n} \pi(m, \tau)\, \pi(n - m, t - \tau),$$

where $0 < \tau < t$. The general solution of the above integral equation can easily be obtained by introducing the probability generating function $g(u, t)$, where

$$(1.3.5) \quad g(u, t) = \sum_{n=0}^{\infty} u^n \pi(n, t) \quad |u| \leqslant 1.$$

From (1.3.4) it follows that $g(u,.)$ for fixed u with $|u| \leqslant 1$ satisfies the relation

(1.3.6) $$g(u, t_1) g(u, t_2) = g(u, t_1 + t_2),$$

so that we have

(1.3.7) $$g(u, t) = \exp [t\phi(u)],$$

where $\phi(.)$ is an arbitrary function of u satisfying the condition

(1.3.8) $$\phi(1) = 0.$$

The function $\phi(.)$ can be determined by specifying the behaviour of $\pi(., t)$ for small t. If we assume that for arbitrary $h > 0$

(1.3.9) $$\pi(n, h) = \lambda_n h + o(h) \quad n \neq 0$$

$$\pi(0, h) = 1 - \sum_{n=1}^{\infty} \lambda_n h + o(h),$$

where $\{\lambda_n\}$ is a positive sequence subject to the condition

(1.3.10) $$\sum_{1}^{\infty} \lambda_n = \lambda < \infty,$$

then it follows that

(1.3.11) $$\phi(u) = -\lambda + \sum_{1}^{\infty} \lambda_n u^n.$$

If we further specialize to the case where

(1.3.12) $$\lambda_1 = \lambda$$

$$\lambda_n = 0 \quad n > 1,$$

we find that

(1.3.13) $$g(u, t) = \exp [-\lambda t(1 - u)],$$

so that

(1.3.14) $$\pi(n, t) = e^{-\lambda t} (\lambda t)^n / n!$$

leading to the Poisson distribution (characterized by the parameter λt) for the number of points in an interval of length t. The stochastic process $Y(t)$ with the specification (1.3.9) and (1.3.12) is called, for obvious reasons, the *simple Poisson process*. The stochastic process $Y(t)$ specified by (1.3.9) is known as the *generalized Poisson process*.

There are several equivalent characterizations of a simple Poisson process. For details the reader is referred to Lindley (1969, Chapter 2) and Parzen (1962, Chapter 4).

1.4 The Yule—Furry process

There are other possible generalizations of the Poisson process apart from the one considered above. If we relax stationarity and deal with the probability mass function $\pi(n, t; m)$, where

(1.4.1) $\quad \pi(n, t; m) = \Pr\{X(t) = n \,|\, X(0) = m\} \quad n \geqslant m,$

we can still regard $\pi(n, . ; m)$ as a function of the single argument t, its dependence on the origin being implicit. Let us first specify the probabilities

(1.4.2) $\quad p(m, h \,|\, n, t) = \Pr\{X(t + h) = m \,|\, X(t) = n\}$

for arbitrarily small $h > 0$ by

(1.4.3) $\quad p(m, h \,|\, n, t) = \lambda_n(t)h + o(h) \qquad m = n + 1$
$$= 1 - \lambda_n(t)h + o(h) \qquad m = n$$
$$= o(h) \qquad\qquad m < n \text{ or } m > n + 1.$$

Thus the Chapman—Kolmogorov relation of type (1.2.4) leads, for small $h > 0$, to

(1.4.4) $\quad \pi(n, t + h, m)$
$$= [1 - \lambda_n(t)h]\,\pi(n, t, m) + \lambda_{n-1}(t)h\pi(n - 1, t, m) + o(h),$$

so that on proceeding to the limit, as $h \to 0$, we have

(1.4.5) $\quad \dfrac{\partial \pi(n, t, m)}{\partial t} = -\lambda_n(t)\,\pi(n, t, m) + \lambda_{n-1}(t)\,\pi(n - 1, t, m).$

If we specialize to the case $\lambda_n(t) = n\lambda$, the above equation can be solved to yield the following explicit solution:

(1.4.6) $\quad \pi(n, t, m) = \dbinom{n - 1}{m - 1} e^{-m\lambda t} (1 - e^{-\lambda t})^{n-m},$

where $n > m \geqslant 1$. This process is known as the "Yule—Furry" process or a "pure birth" process. For, in this case, the process $X(t)$ can be interpreted as the growth of a population whose initial size is m. The probability of a single addition to the population by any one of the members of the population independent of the others at an interval $(t, t + \Delta)$ is $\lambda\Delta$, the probability of multiple births being of a smaller order of magnitude than Δ. Such a model of population growth was first studied by Yule (1924). Furry (1937) also studied and developed the model in connection with the fluctuation problem of cosmic ray showers. Other ramifications of the Poisson process are possible by modifying (1.4.3) in a manner analogous to (1.3.9). A modification which takes into account the risk of death suffered by the population represented by $X(t)$ is given by

$$(1.4.7) \quad \begin{aligned} p(m, h\,|\,n, t) &= n\lambda h + o(h) & m &= n+1 \\ &= n\mu h + o(h) & m &= n-1 \\ &= 1 - n\lambda h - n\mu h + o(h) & m &= n \\ &= o(h) & m &< n-1 \\ & & \text{or } m &> n+1. \end{aligned}$$

The resulting process is of great importance in the theory of population growth. Again, the probabilities (1.4.7) can be modified to admit elements of realism to suit different types of population growth.

2 GENERAL THEORY OF POINT PROCESSES

2.1 Introduction

The theory of stochastic point processes has its origin in attempts by physicists to formulate general problems in Statistical Physics. Yvon (1937) introduced certain correlation functions in connection with the study of an assemblage of molecules distributed over the phase-space. Von Mises (1936) studied the properties of certain functionals associated with random points in the phase-space. Rice (1945) investigated the properties of zero-crossings of certain stochastic processes by introducing probability distribution functions which have turned out to be extremely useful in the general study of point processes. Later, Hermann Wold (1948a, b) introduced in a formal way stationary point processes and studied their properties. D.G. Kendall (1949), Janossy (1950) and Bhabha (1950) investigated some properties of population point processes, with particular reference to the determination of the moments of population size in terms of correlation functions. Ramakrishnan (1950) extended the work of Rice by studying the properties of the distribution functions introduced by the latter. Bartlett (1954, 1966) identified these processes with those considered by Wold and suggested that such processes could be extensively studied from the viewpoint of the general theory of stochastic processes. In fact, current terminology is essentially due to Wold and Bartlett, although special point processes like the Poisson and renewal processes were being investigated as early as the beginning of the century (see, for example, Fry (1928)). Finally, Moyal (1962) combined many of these findings into a formal well-knit theory of point processes and also provided an extension to cover non-Euclidean spaces. A slightly different approach to the theory of point processes was also presented by Harris (1963) in connection with the theory of branching phenomena. In this chapter we give an account of the formal theory of point processes.

2.2 Point processes: preliminary ideas and definitions

Since point processes have been studied by very many individuals with varying background, there have been several definitions of the point process, each appearing quite natural from the viewpoint of the particular problem under study (see, for example, Bartlett (1966, p. 79), Bhabha (1950), Harris (1963), Khintchine (1955) and Wold (1948a)). A comprehensive definition of point process is due to Moyal (1962) who deals with such processes in a general space not necessarily Euclidean, and in this section we shall outline

11

their salient features. Many interesting theorems have been proved by Moyal for general point processes, and in view of their mathematical complexity we have thought best to present only some of these theorems.

Let us consider a set of objects, each of whose state is describable by a point x of a fixed set of points X. Such a collection of objects, which we may call a "population", may be stochastic if there exists a well-defined probability distribution P on some σ-field $\mathbf{B}$ of subsets of the space Φ of all states. We shall assume that members of the population are indistinguishable from one another[(*)]. The state of the population is defined as an unordered set $x^n = \{x_1, x_2, \dots , x_n\}$ representing the situation where the population has n members with one each in the states $x_1, x_2, \dots , x_n$. Thus the population state space Φ is the collection of all such x^n with $n = 0, 1, 2, \dots$, where x^0 denotes the empty population. A *point process* is defined to be the triplet $(\Phi, \mathbf{B}, P)$.

We next observe that it is possible to define measurable functions on Φ since $(\Phi, \mathbf{B})$ is a measure space. Thus we can define a random variable y on the point process $(\Phi, \mathbf{B}, P)$ as a measurable function on Φ, the expected value of y being defined by

$$E\,[y] = \int_{\Phi} y(x^n)\, P(dx^n)$$

$$(2.2.1) \qquad\qquad = \sum_{m=0}^{\infty} \int_{X^m} y_m(x^m)\, P^m(dx^m),$$

where X^m is the m-fold Cartesian product of X with itself. The expected value of y exists if each integral on the r.h.s. of (2.2.1) exists and their sum is absolutely convergent. It is also possible to include the cases when $E\,[y] = \pm \infty$. In section 2.6 we will have occasion to deal with such functions defined on Φ.

Extension to stochastic population processes is achieved by first introducing the probability space $(\Omega, \mathbf{B}_{\Omega}, P_{\Omega})$. Let $(X_t, \mathbf{B}_t; t \in T)$ be an indexed family of individual measure spaces and $(\Phi_t, \mathbf{B}_t; t \in T)$ the associated family of population measure spaces. Now for each $t \in T$, we can define a random variable $x^n(t, .)$ by a mapping $\Omega \to \Phi_t$. If $\Phi_T = \Pi_{t \in T} \Phi_t$ and $\mathbf{B}_T = \Pi_{t \in T} \mathbf{B}_t$, then it is clear that $\mathbf{B}_T$ is the σ-field generated by the field of all measureable cylinders $A_k \times \Phi_{T-k}$, where k is a finite subset of T and $A_k \in \mathbf{B}_k = \Pi_{t \in k} \mathbf{B}_t$. Thus the transformation $\omega \to x^n (., \omega)\,(\omega \in \Omega)$ by construction is measurable and yields a probability distribution P_T on $\mathbf{B}_T$, so that we can conclude that $(\Phi_T, \mathbf{B}_T, P_T)$ is a population point process. If T is finite, we indeed have a multivariate population point process. As an example let us consider the sequence of population processes $x_1, x_2, \dots , x_n \dots$ with individual measure spaces respectively $(X_1, \mathbf{B}_{x_1}), (X_2, \mathbf{B}_{x_2}), \dots , (X_n, \mathbf{B}_{x_n}) \dots$ and

---

(*) Moyal (1962) has also considered the case of distinguishable members.

associated population measure spaces $(\Phi_1, \mathbf{B}_{x_1}), (\Phi_2, \mathbf{B}_{x_2}), \ldots, (\Phi_n, \mathbf{B}_{x_n}) \ldots$.
Let us assume that the sequence of point distributions corresponds to the objects in different generations, each object of the ith generation being describable by the value $x_i \in X_i$ that it has at its birth. If we assume that the quantity x_i completely characterizes the object, we can conclude that a knowledge of the point process x_n is sufficient to "predict" the future and that the additional knowledge of the point distribution corresponding to the previous generations is irrelevant. Thus the sequence x_i is a generalized Markov process whose states are not members but are themselves point processes. In fact, in branching phenomena, we usually have an additional simplifying feature corresponding to the independence of objects in their ability to procreate. For an interesting development of the theory along these lines, the reader is referred to the work of Harris (1963). In passing we note that the concepts and properties of stochastic independence, conditional distributions and conditional expectations carry over *in toto* to such stochastic population point processes.

2.3 Counting measures and product densities

There is an alternative and more purposeful method of characterizing a stochastic population by assigning to sets A in the individual state space X the number of objects or individuals $N(A)$ each of whose state x belongs to A $(x \in A)$. We shall first confine our attention to populations whose total size is finite with probability one; later we shall remove these restrictions. To fix ideas, let us consider the special case when the population consists of n individuals in the states $x_1, x_2, \ldots, x_n$. The number of individuals with states in a given arbitrary $A \subset X$ is given by

(2.3.1) $$N(A \,|\, x^n) = \sum_{i=1}^{n} \delta(A \,|\, x_i),$$

where $\delta(A \,|\, .)$ is the characteristic function of the set A defined by

(2.3.2) $$\begin{cases} \delta(A \,|\, x) = 1 & x \in A, \\ = 0 & \text{otherwise.} \end{cases}$$

We note that for each fixed $A \in \mathbf{U}$, the class of all subsets of X, $N(A \,|\, .)$ is a function defined on the population state space (which is symmetric), while for fixed $x^n \in \Phi$, $N(. \,|\, x^n)$ is a function defined on $\mathbf{U}$ and having the following properties:

(i) non-negative and finite;
(ii) integral valued;
(iii) completely additive in the sense that if $\{A_t, t \in T\}$ is any arbitrary collection of mutually disjoint sets in X, then there can be at most a finite number of sets $A_{t_1}, A_{t_2}, \ldots A_{t_m}$ such that

$$N_{A_t} \geqslant 1,$$

(2.3.3) $$N\left(\sum_{t \in T} A_t\right) = \sum_{i=1}^{m} N(A_{t_i}).$$

Thus we have a counting measure $\mathfrak{N}$ on the set $\mathbf{U}$ of all subsets of X. In other words, the point process $(\Phi, \mathbf{B}, P)$ generates a probability space $(\mathfrak{N}, \mathbf{B}_N, P_N)$ which we shall call a *counting process*. It can also be proved that every such *counting process* determines a point process (see Moyal (1962)), so that there is a one-to-one correspondence between Φ and $\mathfrak{N}$.

We next note that for fixed A, the function $N(A \mid .)$ on Φ is measurable if and only if A is measurable (see problem 2.1, below). If $\{A_1, A_2, \dots, A_l\}$ is a finite measurable partition of X and $k_1 + k_2 + \dots + k_l = n$, we have

$$P_N\{N(A_i) = k_i, i = 1, 2, \dots, l\}$$

(2.3.4) $$= \frac{n!}{k_1! \, k_2! \dots k_l!} P_S^{(n)}(A_1^{k_1} \times A_2^{k_2} \dots \times A_l^{k_l}),$$

the suffix S reinforcing the fact that the distribution on $\mathbf{B}$ is symmetric. We can use this result to deduce the joint distributions of $N(A)$ for any finite collection of measurable sets in X in terms of the distribution P_S on $\mathbf{B}$. For instance, the distribution of $N(A)$ is given by

(2.3.5) $$P_N(N(A) = n) = \sum_{k=0}^{\infty} P_N(N(A) = n, N(X - A) = k)$$

or

(2.3.6) $$P_N(N(A) = n) = \sum_{k=0}^{\infty} \binom{n+k}{n} P_S^{(n+k)}(A^n \times (X - A)^k),$$

while the mean value of $N(A)$ is given by

$$E[N(A)] = \sum_{n=0}^{\infty} n P_N(N(A) = n)$$

(2.3.7) $$= \sum_{n=0}^{\infty} n \sum_{k=n}^{\infty} \binom{k}{n} P_S^{(k)}(A^n \times (X - A)^{k-n}).$$

Using (2.2.6), we find

(2.3.8) $$E[N(A)] = \sum_{n=1}^{\infty} n P_S^{(n)}(A \times X^{n-1}).$$

We can interpret equation (2.3.8) as the assignment of the value at A of a measure $M(A)$ defined on $\mathbf{B}$. We further note that the measure is finite if and only if the mean value is finite, and σ-finite if and only if $M(A_i)$ is finite for each A_i of some measurable countable partition $\{A_i\}$ of X.

We can deal with the higher moments of the number of individuals distributed over any given arbitrary set $A \subset X$. This is facilitated by the introduction of the kth product measure N_k generated by N on $\mathbf{B}^k$. Thus it follows from (2.3.1) for each $A^k \in \mathbf{B}^k$ that

$$(2.3.9) \quad N_k(A^k | x^n) = \sum_{i_1=1}^{n} \sum_{i_2=1}^{n} \dots \sum_{i_k=1}^{n} \delta(A^k | x_{i_1}, x_{i_2}, \dots, x_{i_k}),$$

where $\delta(. | x_{i_1}, x_{i_2}, \dots, x_{i_k})$ is the product measure

$$\delta(. | x_{i_1}) \times \delta(. | x_{i_2}) \times \dots \times \delta(. | x_{i_k}).$$

The function $N_k(. | x^n)$ assigns the measure 1 to the singleton element $\{(x_{i_1}, x_{i_2}, \dots, x_{i_k})\}$ in X^k, where $x_{i_j} \in \{x_1, x_2, \dots, x_n\}\, j = 1, 2, \dots, k$ and 0 to the rest of X^k. Thus $N_k(. | x^n)$ is itself a counting measure on $\mathbf{B}^k$ in just the same way as $N(. | x^n)$ is on $\mathbf{B}$. The expected value of N_k, which we denote by M_k, can be identified as the measure on $\mathbf{B}^k$ which is again finite if m_k, the kth moment of the counting process $\mathbf{N}$, is finite and σ-finite if $M_k(A_i^k)$ is finite for each set A_i^k of some measurable countable partition of X.

Explicit expression of M_k in terms of P_S by the use of higher-moment measures is fairly complex. Such expressions contain concentrations along subsets of lower dimensions. For instance, M_2 has a concentration along the diagonal corresponding to the pairs (x_i, x_i):

$$(2.3.10) \quad M_2(A^2) = M_{(2)}(A^2) + M(DA^2)$$

where

$$(2.3.11) \quad M_{(2)}(A^2) = \mathrm{E}\,[N_{(2)}(A^2)]$$

$$(2.3.12) \quad N_{(2)}(A^2) = \sum_{\substack{i_1=1 \\ i_1 \neq i_2}}^{n} \sum_{i_2=1}^{n} \delta(A^2 | x_{i_1}, x_{i_2}) \quad (n \geqslant 2)$$

and DA^2 is the set of all $x \in X$ such that $x = y$ and $(x, y) \in A^2$. We next observe that in the definition of $N_k(. | x^n)$, we assign 1 to the counting measure if the singleton element $\{(x_{i_1}, x_{i_2}, \dots, x_{i_k})\} \in X^k$. This does not cover the most general case. For instance, the population state is defined as the unordered set $\{x_1, x_2, \dots, x_n\}$. If some of the x_i's are equal (this corresponds to non-regular or unorderly processes in the notation of Khintchine (1955)), then the counting measure must be modified suitably. For the present let us confine our attention to regular or orderly processes, In such a case $M_{(2)}(A^2)$ can be interpreted as the second factorial moment. The general equation $M_k = \mathrm{E}\,[N_k]$ can be expressed in terms of the factorial moments by introducing the counting measure $N_{(k)}$ on $\mathbf{B}^k$ by the relation

$$(2.3.13) \quad N_{(k)}(A^k | x^n) = \sum_{i_1 \neq i_2 \neq \dots \neq i_k} \delta(A^{(k)} | x_{i_1}, x_{i_2}, \dots, x_{i_k}) \, (n \geqslant k).$$

The above equation shows that $N_{(k)}(. | x^n)$ assigns measure one to every singleton $\{(x_{i_1}, x_{i_2}, \dots, x_{i_k})\}$, such that $x_{i_1} \neq x_{i_2} \neq \dots \neq x_{i_k}$ and

$x_{i_j} \in \{x_1, x_2,..., x_n\} j = 1, 2,..., k$, and zero to the rest of the points of X^k.
It is clear that the expectation of $N_{(k)}(A^k|.)$, which we denote by $M_{(k)}$, is
the mean distribution of $N_{(k)}$, having properties very similar to M_k. The
expectation of $N_{(k)}$ is given by

$$M_{(k)}(A^k) = \sum_{n=k}^{\infty} \int_{X^n} \sum_{i_1 \neq ... \neq i_k} \delta(A^k|x_{i_1}, x_{i_2},..., x_{i_k}) P_S^{(n)} dx_n$$

$$(2.3.14) \qquad = \sum_{n=k}^{\infty} \frac{n!}{(n-k)!} P_S^{(n)} (A^k \times X^{n-k}).$$

Next we note that $N_k(A^k|.)$ is related to $N_{(k)}(A^k|.)$ by

$$(2.3.15) \quad N_k(A^k|.) = \sum_{l=1}^{k} C_l^k N_{(l)}(A^l|.),$$

where C_l^k are the Stirling numbers of the second kind representing the
$(k-1)$-fold degeneracy in a product of k terms due to coalescence. With
the help of (2.3.15) we find

$$(2.3.16) \qquad M_k(A^k) = \Sigma C_l^k M_{(l)}(A^l).$$

Let μ be a measure on $\mathbf{B}$ and let μ_n be the nth product measure generated by μ on $\mathbf{B}^n$. If $P_S^{(n)}(.)$ is absolutely continuous with respect to the
product measure μ_n, then $M_{(k)}$, if it exists, is absolutely continuous with
respect to μ_k and its density $h_{(k)}(x^k)$ is given by

$$(2.3.17) \quad h_{(k)}(x^k)$$

$$= \sum_{n=k}^{\infty} \frac{n!}{(n-k)!} \int_{X^{n-k}} \pi_n(x_1, x_2,..., x_n) \mu_{n-k}(dx_{k+1} ... dx_n),$$

where π_n is the density of $P_S^{(n)}$ with respect to μ_n. The above relations
(2.3.16) and (2.3.17) were derived by Bhabha (1950) where X is the real
line. The density functions $h_{(k)}(.)$ were introduced by Rice (1945) and
Ramakrishnan (1950) and are known in the literature as *product densities*.
The more general formulation for an arbitrary set X is of course due to
Moyal (1962). Janossy (1950) studied the functions $h_{(k)}(.)$ indirectly by
dealing with the right-hand side of (2.3.17).

So far the discussion on point processes has been confined to populations whose total size is finite with probability one. We now briefly indicate how the theory can be extended to cover the case when the total
population size can be infinite with non-zero probability. This is best done
by introducing the set of all positive integral-valued completely additive
measures N_ω on $\mathbf{B}$, such that $N_\omega(X_k) < \infty$ for each X_k of a given *fixed*
monotonic increasing sequence of measurable sets $\{X_i: X_i \uparrow X\}$. If n_ω is
the set, a σ-finite counting process is the probability space $\mathbf{N}_\omega = (\mathrm{n}_\omega, \mathbf{B}_\omega, P_\omega)$,
where $\mathbf{B}_\omega$ is a suitably defined σ-field of sets in n_ω and P_ω is a probability

distribution on $\mathbf{B}_\omega$. Given such a probability space, consider n_k, the set of all restrictions of elements n_ω to X_k. Thus we can arrive at a finite counting process $\mathbf{N}_k = (n_k, \mathbf{B}_k, Q_k)$, where $\mathbf{B}_k$ is the class of sets S_k in n_k whose continuation S_k^ω to X belongs to $\mathbf{B}_\omega$, and Q_k is the function on $\mathbf{B}_k$ whose value at S_k is given by $Q_k(S_k) = P_\omega(S_k^\omega)$. The finite counting process is thus the restriction of $\mathbf{N}_\omega$ under the set X_k. It can be shown that the sequence $\mathbf{N}_k$ of finite counting processes determines a unique $\mathbf{N}_\omega$ such that $\mathbf{N}_k$ is the restriction of $\mathbf{N}_\omega$ to the counting process corresponding to the set X_k, $k = 1, 2, \ldots$.

We next note that the definitions of moment and factorial moment distributions given in (2.3.14), (2.3.15), (2.3.16) and (2.3.17) extend to the σ-finite case. If the restrictions $\mathbf{N}_k$ of $\mathbf{N}_\omega$ corresponding to the set X_k have finite mean distributions $M^{(k)} = \mathrm{E}\,[N_k]$, then the mean distribution $M = \mathrm{E}\,[N_\omega]$ is a σ-finite measure on $\mathbf{B}$ such that $M(X_k) = M^{(k)}(X_k) < \infty$, $k = 1, 2, \ldots$. Exactly similar considerations apply to the higher-order product measures.

Extension of the above ideas to cover the stochastic population point processes $(\Phi_T, \mathbf{B}_T, P_T)$ introduced in section 2.2 is possible, although it involves a lot of technical details, particularly in the case when T is infinite. An outline of the method of extension to such processes will be found in Moyal's work. For each $t \in T$ and $\omega \in \Omega$, the transformation (2.3.1) defines a counting measure $N(.,\omega)$ on $\mathbf{B}_t$. Thus if $(n_t, \mathbf{B}_{N_t}; t \in T)$ is the family of counting measures generated, then the transformation (2.3.1) yields the stochastic population point process $(n_T, \mathbf{B}_{N_T}, P_{N_T})$ in the counting process formulation introduced earlier with $n_T = \Pi_{t\in T} n_t$ and $\mathbf{B}_{N_T} = \Pi_{t\in T} \mathbf{B}_{N_t}$. The derivations leading through (2.3.14) to (2.3.17) follow, except that M_k, h_k and π_n are functions of $t \subset T$ as well. The functions h_k coincide with the product densities introduced by Bhabha (1950) and Ramakrishnan (1950) for the study of electromagnetic cascades, with t standing for the thickness of the material traversed by the particles. Bhabha and Ramakrishnan (1950) studied the two *expectation processes* generated by the first- and second-order product densities and were able to make estimates of the shower size with the help of the properties of Markov branching processes, which will be introduced in section 2.6.

2.4 Generating functionals

The concept of the probability generating functional was first explicitly introduced by D.G. Kendall (1949) in the context of population growth. Its usefulness in the study of general stochastic processes with special reference to regenerative and branching processes was demonstrated by Bartlett and D.G. Kendall (1951).

As before, let $(\Phi, \mathbf{B})$ be the point-process measure space generated by the measure space $(X, \mathbf{B})$ and let $\mathfrak{M}$ be the linear vector space of all bounded measurable complex-valued functions ξ defined on X. The linear space $\mathfrak{M}$ becomes a Banach space under the usual norm

(2.4.1)
$$\|\xi\| = \sup_{x \in X} |\xi(x)|.$$

The probability generating functional (p.g.fl) of a probability distribution on **B** is defined as the expectation value of the symmetric measurable function

$$w(x^n) = \xi(x_1)\xi(x_2) \dots \xi(x_n) \text{ on } \Phi:$$

(2.4.2) $G[\xi] = E\{w\} = \sum_{n=0}^{\infty} \int_{X^n} \xi(x_1)\xi(x_2) \dots \xi(x_n) P_S^n \, (dx^n).$

Writing

(2.4.3)
$$p_n = p^n(X^n),$$

we observe that $G[\xi]$ is a functional defined on a domain $D_G \subset \mathfrak{M}$ which includes the sphere $S_g = \{\xi: \|\xi\| \leqslant r\}$, where r ($\geqslant 1$) is the radius of convergence of the series $\sum_{n=0}^{\infty} p_n z^n$.

It is easy to note that if we define the function $\xi(.)$ by

(2.4.4)
$$\xi(x) = z \quad x \in A \subset X$$
$$= 0 \quad \text{otherwise},$$

where z is a complex number, $G[\xi]$ is the generating function (p.g.f.) of the probability distribution of the population state over the set A. To obtain the multivariate p.g.f. corresponding to a finite measurable partition $\{A_1, A_2 \dots A_k\}$ of X, we define

(2.4.5) $g(z_1, A_1; z_2, A_2; \dots ; z_k, A_k) = E\{\sum_{i=1}^{k} z_i^{N(A_i)}\},$

where each z_i is a complex variable of modulus $\leqslant 1$. If we define $\xi(.)$ by

(2.4.6)
$$\xi(x) = z_i \quad x \in A_i,$$

it follows from (2.4.2) that

(2.4.7) $G[\xi] = \sum_{n_1=0}^{\infty} \sum_{n_2=0}^{\infty} \dots \sum_{n_k=0}^{\infty} \dfrac{(n_1 + n_2 + \dots + n_k)!}{n_1! \, n_2! \dots n_k!}$
$$\times z_1^{n_1} z_2^{n_2} \dots z_k^{n_k} P_S^{(n_1 + n_2 + \dots \, n_k)}(A_1^{n_1} \times A_2^{n_2} \times \dots \times A_k^{n_k}).$$

By identifying the coefficients of $z_1^{n_1} z_2^{n_2} \dots z_k^{n_k}$ in (2.4.7), we find an explicit expression for the probabilities $P_N(N(A_i) = n_i, i = 1, 2, \dots, k)$.

We next observe that the factorial moment distributions $M_{(k)}$, whenever they exist as σ-finite distributions on $\mathbf{B}^k$, can be generated from $G[.]$. Let η, ξ be fixed elements of $\mathfrak{m}$ and let $\|\eta\| < 1$. If r is the largest real number such that $\eta + \lambda\xi \in S_g$ for $|\lambda| < r$, $G[\eta + \lambda\xi]$, considered as a function of λ

in the complex plane Λ, can be written as

$$(2.4.8) \quad G[\eta + \lambda\xi] = \sum_{n=0}^{\infty} \sum_{k=0}^{n} \binom{n}{k} \lambda^k \int_{X^n} \xi(x_1)\xi(x_2) \dots \xi(x_k)\eta(x_{k+1}) \dots \eta(x_n)$$
$$P_S^{(n)}(dx^n),$$

so that we can conclude that $G[.]$ is an analytic function of λ in any closed region interior to the circle $|\lambda| < r$. Rearranging the series in the form

$$(2.4.9) \quad G[\eta + \lambda\xi] = \sum_{k=0}^{\infty} \lambda^k \sum_{n=k}^{\infty} \int_{X^n} \xi(x_1)\xi(x_2) \dots \xi(x_k)\eta(x_{k+1}) \dots \eta(x_n)$$
$$P_S^n(dx^n),$$

we note that $G[\eta + \sum_{i=1}^{k} \lambda_i \xi_i]$ $(\|\eta\| < 1, \xi_i \in \mathfrak{m}, i = 1, 2, \dots, k)$, has first partial derivatives to the λ_i's in some open region D_0 of Λ^k containing the origin and is therefore an analytic function of k-complex variables $\lambda_1, \lambda_2, \dots, \lambda_k$ in D_0. Thus we can define the "kth order variation of G" (following Hille and Phillips (1957)) as

$$\delta^k_{\xi_1 \xi_2 \dots \xi_k} G[\eta] = \left\{ \frac{\partial^k}{\partial\lambda_1 \partial\lambda_2 \dots \partial\lambda_k} G\left[\eta + \sum_{i=1}^{k} \lambda_i\xi_i\right] \right\}$$
$$\lambda_1 = \lambda_2 = \dots = \lambda_k = 0$$
$$= \sum_{n=k}^{\infty} \frac{n!}{(n-k)!} \int_{X^n} \xi_1(x_1) \xi_2(x_2) \dots \xi_k(x_k)$$
$$(2.4.10) \qquad\qquad \times \eta(x_{k+1}) \dots \eta(x_n)P_S^n(dx^n).$$

If we proceed to the limit $\eta \to 1$, we find from (2.3.14)

$$(2.4.11) \quad \lim_{\eta \to 1} \delta^k_{\xi_1 \xi_2 \dots \xi_k} G[\eta] = \int_{X^k} \xi_1(x_1)\xi_2(x_2) \dots \xi_k(x_k)M_{(k)}(dx^k).$$

If we assume that $M_{(k)}$ is absolutely continuous with respect to the kth product measure on $\mathbf{B}^k$, then its density $h_k(x^k)$ can be obtained by choosing $\xi_i(x)$ to correspond to the Dirac delta function $\delta(x - \bar{x}_i)$. Thus the functional derivatives of the probability generating functional, evaluated at appropriate points, can be identified with the product densities.

We can also obtain another parallel simple result for the moment distribution by setting $\eta = 0$.

$$(2.4.12) \quad \delta^k_{\xi_1 \xi_2 \dots \xi_k} G[0] = k! \int_{X^k} \xi_1(x_1)\xi_2(x_2) \dots \xi_k(x_k)P_S^{(k)}(dx^k).$$

If we set

$$\xi_i(x) = \delta(A_i|x) \quad i = 1, 2, \dots, k,$$

where $A_1, A_2,..., A_k$ are arbitrary measurable sets in X, then we find

$$(2.4.13) \quad P_S^{(k)}(A_1 \times A_2 \times ... \times A_k) = \frac{1}{k!} \delta_{\xi_1 \xi_2 ... \xi_k}^k G[0],$$

a result proving that G determines P_S uniquely, so that there is a one-to-one correspondence between the class of all symmetric probability distributions on $\mathbf{B}$ and the class of all probability generating functionals on the unit sphere S_0 in $\mathfrak{M}$.

We next observe that if $r_g > 1$, then all $M_{(k)}$ are finite, and we can set $\eta = 1$ in (2.4.8) to obtain the expansion of G in terms of factorial moment distributions:

$$(2.4.14) \quad G[1 + \lambda\xi] = 1 + \sum_{k=1}^{\infty} \lambda^k \int_{X^k} \xi(x_1)\xi(x_2) ... \xi(x_k)M_{(k)}(dx^k).$$

Now we can substitute $\lambda\xi - 1$ for $\lambda\xi$ in the left-hand side of (2.4.14) and we find

$$(2.4.15) \quad G[\lambda\xi] = \sum_{n=0}^{\infty} \lambda^n \sum_{k=n}^{\infty} \frac{(-)^{k-n}}{n!\,(k-n)!} \int_{X^n} \xi(x_1)\xi(x_2) ... \xi(x_k)M_{(k)}(dx^k).$$

If A is any measurable subset of X, we can choose

$$(2.4.16) \qquad\qquad \xi(x) = \delta(A|x)$$

Comparing (2.4.15) with (2.4.2), we find

$$(2.4.17) \quad P_S^{(n)}(A^n) = \sum_{k=n}^{\infty} \frac{(-1)^{k-n}}{n!\,(k-n)!} M_{(k)}(A^{(n)} \times X^{k-n}),$$

the inverse of relation (2.3.14). If the corresponding densities are defined, we have the inverse formula (due to Kuznetsov and Stratonovich (1956)) corresponding to (2.3.17):

$$(2.4.18) \quad \pi_n(x_1, x_2,..., x_n)$$

$$= \sum_{k=n}^{\infty} \frac{(-)^{k-n}}{n!\,(k-n)!} \int_{X^{k-n}} h_k(x_1, x_2,..., x_k)\mu_{k-n}\,(dx_{n+1}\,dx_{n+2}... dx_k).$$

Finally we proceed to connect the characteristic functional of the point process with the generating functional defined by (2.4.2). The idea of the characteristic functional is due to Bochner (1947) and Le Cam (1947). Bochner introduced an overall characteristic function $\phi[h]$ taking numerical values and defined over the elements of a certain vector space, whilst Le Cam was concerned with a physical problem connecting river water flow with rainfall. A complete mathematical account of the characteristic functional was later given by Bochner (1955) in his monograph. For the point process defined in this chapter, the characteristic functional may be defined by

$$(2.4.19) \quad \Phi[\theta] = E\left[\exp i \int_X \theta(x)N(dx)\right]$$

$$= \sum_{n=0}^{\infty} \int_{X^n} \exp i \left[\theta(x_1) + \theta(x_2) + \dots + \theta(x_n)\right] P_S^{(n)}(dx^n),$$

so that we can obtain $\Phi[\theta]$ from the generating functional $G[\xi]$ by the substitution $\xi = e^{i\theta}$, where θ is a real-valued function defined on X. Thus $\Phi[\theta]$ is a functional on the whole space $\mathfrak{m}_k$ of all real-valued bounded measurable functions on X. When the moment distributions exist, we can use the characteristic functional to generate them (see problem 2.3, below). The functional derivatives of the characteristic functional coincide with the product densities whenever the latter are defined.

2.5 Point processes on the real line: product densities and their extensions

We shall now specialize to the case when X is the real line. The members of the population will be called *events* and the subsets A_k can be assumed to be intervals. The counting measure $N(.)$ is an integral-valued countably additive σ-finite measure on the real line. Every such measure is uniquely associated with a sequence of points $\{t_i\}$ on the real line by the formula

$$N(A) = \text{number of members of the sequence } \{t_i : t_i \in A\}.$$

If we take expectation of the counting measure, we obtain the first moment measure. In what follows it will be assumed that it is always a Borel measure. This assumption implies that the expected number of events in every finite A is finite, which in turn implies that with probability one there are only a finite number of events in a finite interval.

We shall use the following notation:

$N(t, x)$: the random variable representing the number of events in the interval $(t, t + x)$;

$d_x N(t, x)$: the random variable representing the number of events in the interval $(t + x, t + x + dx)$;

$$(2.5.1) \qquad p(n; t, x) = \Pr\{N(t, x) = n\}.$$

We next introduce the notion of *regularity* or *orderliness*. A point process defined on the real line is said to be *orderly* or *regular* if the probability governing the counting measure satisfies the condition

$$(2.5.2) \quad \sum_{n \geqslant 2} p(n; t, t + \Delta) = o(\Delta) \quad \text{for small } \Delta.$$

The notion of orderliness was introduced by Khintchine (1955) in connection with problems of queuing theory where the events correspond to arrivals. The basic idea behind the notion is to introduce some kind of order among the arrivals and this is possible only if (2.5.2) is satisfied. In fact, condition (2.5.2) was imposed on point processes by D.G. Kendall (1949) and Ramakrishnan (1950). An interesting consequence of (2.5.2) observed by both

Kendall and Ramakrishnan is given by the relation

(2.5.3) $\quad \lim_{\Delta \to 0} \mathrm{E}\{N(t, \Delta)\}/\Delta = \lim_{\Delta \to 0} \mathrm{Pr}\{N(t, \Delta) \geq 1\}/\Delta,$

so that we can conclude that

(2.5.4) $\quad \mathrm{E}\{N(t, \Delta)\} = \mathrm{var}\{N(t, \Delta)\}$ for small Δ,

and hence that the random variable $N(t, \Delta)$ is asymptotically Poisson in character. Bhabha (1950) observed the same result and proved that although the counting variables assume a Poisson character for small Δ, they do exhibit considerable deviation from Poisson nature for finite Δ.

The densities of the factorial moment measures $M_{(k)}$ introduced in the previous section can be conveniently defined by

(2.5.5) $\quad h_n(x_1, x_2,..., x_n)$

$$= \lim_{\Delta_1, \Delta_2,...\Delta_n \to 0} \mathrm{E}\left[\prod_{i=1}^{n} N(x_i, \Delta_i)\right]/\Delta_1 \Delta_2 ... \Delta_n \quad (x_1 \neq x_2 \neq ... \neq x_n.$$

or equivalently by

(2.5.6) $\quad h_n(x_1, x_2,... x_n)$

$$= \lim_{\Delta_1, \Delta_2...\Delta_n \to 0} [\mathrm{Pr}\{N(x_i, \Delta_i) \geq 1 \quad i = 1, 2,..., n\}/\Delta_1 \Delta_2 ...\Delta_n]$$

$$(x_1 \neq x_2 \neq ... \neq x_n)$$

It should be noted that the equivalence of (2.5.5) and (2.5.6) is a consequence of regularity. When regularity is relaxed, (2.5.5) and (2.5.6) do differ. This point will be further discussed below. Since $h_n(.)$ is a product of the density of expectation measures at different points, the density is aptly called the *product density* (see Ramakrishnan (1950)). Rice (1945) had made explicit use of these density functions in dealing with events corresponding to the zero crossings of a Gaussian random process. In many processes it is convenient to deal with the product densities directly, and we shall illustrate the point in the next section when we give some examples. The second-order product density has a central role in stationary processes since its Fourier transform yields the spectral density of the counting process.

If we relax the regularity condition, a number of difficulties arise. As has been observed in 2.3, the factorial-moment measures contain concentrations along subsets of lower dimensions. In fact, the use of the factorial-moment measure is primarily for avoiding such situations, since normal moment measures do contain such concentrations. Thus we have to admit delta functions into the corresponding factorial moment densities. For instance, a multiple event of order $m(i)$ occurring at x_i will be represented by the density function in which the argument repeats $m(i)$ times, so that the corresponding density function will contain a product of $m(i)$ delta functions of the form $\prod_{j=i+1}^{i+m\,(i)} \delta(x_j - \sigma x_i)$. This poses a piquant situation as compared with the regular case where, although the product density, say $h_n(x_1, x_2,..., x_n)$, is defined for $x_1 \neq x_2 \neq ... \neq x_n$ by (2.5.5), the value of the density $h_n(x_1, x_2, ..., x_n)$ when two of its arguments x_i, x_j become equal can be defined, by virtue of its continuous nature, as being the limit as $x_i \to x_j$. This point will be further discussed in Chapter 4 when we deal with stationary point processes. An alternative approach consists in visualizing the point process as composed of locations of events of the point process, the locations themselves constituting a regular point process and the multiplicities (of events) being assigned to these locations. This device is due to Srinivasan (1961) who constructed multiple product densities on this basis by using (2.3.17). In such a case we can always start with the probability density function

$$\pi_{[n_1,n_2,...,n_m]}(x_1, x_2,..., x_m)$$

of a complex $[n_1, n_2,..., n_m]$ corresponding to the multiple events of weights $n_1, n_2,..., n_m$ respectively at $(x_1, x_1 + dx_1) ... (x_m, x_m + dx_m)$. A formal construction of these density functions starting from the basic probability space $(\Phi, \mathbf{B}, P)$ can be made by modifying the counting measure (2.3.1).

We can introduce the random variable $N_i(t, x)$ corresponding to the number of distinct events of multiplicity i in $(t, t + x)$, and equations (2.5.3) and (2.5.4) hold good for such random variables. In this case a product density of degree n will be defined by

$$(2.5.7) \quad h_{n[n_1,n_2,...,n_m]}(x_1, x_2,..., x_n)$$
$$= \lim_{\Delta_1, \Delta_2...\Delta_n \to 0} \mathrm{E} \left[\prod_{i=1}^{n} N_{m_i}(x_i, \Delta_i) \right]/\Delta_1 \Delta_2 ...\Delta_n \quad x_1 \neq x_2 \neq ... \neq x_n.$$

The moment formula (2.3.16), valid in this case, describes only the number of locations and throws no light on the distribution of the total number of events. If $N(t, x)$ is the random variable corresponding to the number of events (not locations) in $(t, t + x)$ then it is clear that

$$(2.5.8) \qquad\qquad N(t, x) = \sum_i i N_i(t, x).$$

Obviously if there is no upper limit to the multiplicity, questions of convergence arise, and the sum on the right-hand side of (2.5.8) therefore needs to be justified anew each time a new stochastic point process is generated. With the help of (2.5.8) a general moment formula can be obtained. For simplicity, let us consider the case when $i \leqslant 2$. Using (2.5.8) and (2.3.16), we find that

$$(2.5.9) \quad \mathrm{E}\,[N(t, x)] = \int_t^{t+x} h_{1[1]}(u)\,du + 2 \int_t^{t+x} h_{1[2]}(u)\,du$$

$$(2.5.10) \quad E\,[N(t,x)]^2 \;=\; \int_t^{t+x} h_{1[1]}(u)\,du \;+\; 4\int_t^{t+x} h_{1[2]}(u)\,du$$

$$+\;2\int_t^{t+x}\int_t^{t+x} h_{2[1,2]}(u,v)\,du\,dv \;+\; 4\int_t^{t+x}\int_t^{t+x} h_{2[2,2]}(u,v)\,du\,dv.$$

The moment formulae for the most general case can be derived. For details the reader is referred to Srinivasan (1961, 1969) (see also problem 2.5, below).

2.6 Some examples

We shall now illustrate the general theory by examples.

(i) *Inhomogeneous Poisson process*   With reference to the origin (fixed) let us consider the events only in the interval $(0, T)$. The Poisson process is characterized by the independence of events. Let

$$(2.6.1) \qquad\qquad \lim_{\Delta \to 0} \frac{\Pr\{N(t,\Delta)=1\}}{\Delta} \;=\; \lambda(t).$$

Then we find, using independence of events,

$$(2.6.2) \quad \pi_n(x_1, x_2,\dots, x_n) \;=\; \lambda(x_1)\lambda(x_2)\dots\lambda(x_n)\exp - \int_0^T \lambda(u)\,du$$

$$(x_i \in (0,T),\; i = 1, 2,\dots, n)$$

$$(2.6.3) \quad h_n(x_1, x_2,\dots, x_n) \;=\; \lambda(x_1)\lambda(x_2)\dots\lambda(x_n).$$

The characteristic functional $\Phi(\theta, T)$ is given by

$$\Phi(\theta, T) \;=\; E\,\{\exp i \int_0^T \theta(t)\,d_t N(0,t)\}$$

$$=\; E\,\{\lim_{\substack{k \to \infty \\ k\Delta = T}} \prod_{j=1}^k \int_{(j-1)\Delta}^{j\Delta} \theta(t)\,d_t N(0,t)\}$$

$$=\; \lim_{\substack{k \to \infty \\ k\Delta = T}} \prod_{j=1}^k E\,\{\int_{(j-1)\Delta}^{j\Delta} \theta(t)\,d_t N(0,t)\}$$

$$=\; \lim_{\substack{k \to \infty \\ k\Delta = T}} \prod_{j=1}^k \{(1 - \lambda(t_j)\Delta) + \lambda(t_j)\Delta e^{i\theta(t_j)}\}$$

$$(2.6.4) \qquad\qquad =\; \exp \int_0^T \lambda(t)\,(e^{i\theta(t)} - 1)\,dt.$$

An irregular point process can be generated by first considering Poisson locations and then assigning multiple events to each location, the multiplicities at different locations being independently and identically distributed.

This process will be discussed in detail in Chapter 4 when we deal with stationary point processes.

(ii) *Inhomogeneous Yule–Furry process*   Let us consider an inhomogeneous birth process generated by a single individual (called "primary") at time $t = 0$. The primary and its secondaries generate a population in accordance with the following assumptions:

(1) the subpopulations generated by two coexisting individuals develop in complete independence of one another;
(2) an individual present at time t has a probability

$$\lambda(t)\Delta + o(\Delta)$$

of producing another in the time-interval $(t, t + \Delta)$;
(3) the probability that an individual present at time t gives birth to more than one individual in the interval $(t, t + \Delta)$ is $o(\Delta)$.

The population process is known as the Yule–Furry process (Yule (1924), Furry (1937)). The events of the point process correspond to the births occurring at various epochs of the time-axis. We can confine our attention to the fixed interval $(0, T)$. In view of assumption (3) the population is finite over the interval $(0, T)$ with probability one.

The function $\pi_n(t_1, t_2,..., t_n)$ $(t_i \in (0, T), i = 1, 2,..., n)$ introduced in (2.3.11) has the following interpretation:

$$(2.6.5) \quad \pi_n(t_1, t_2,..., t_n) = \lim_{\Delta_1, \Delta_2,..., \Delta_n \to 0} \Pr \{N(t_i, \Delta_i) = 1, N(0, t_i) = 0,$$

$$N(t_i + \Delta_i, t_{i+1}) = 0 \quad i = 1, 2,..., n\}/\Delta_1 \Delta_2 ... \Delta_n, \quad (t_{n+1} = T).$$

We note that the process is not stationary, even if the probabilities $\lambda(.)$ are assumed to be constants. However, using assumptions (1) to (3), we find that

$$(2.6.6) \quad \pi_n(t_1, t_2,..., t_n) = n! \, \lambda(t_1)\lambda(t_2) ... \lambda(t_n)$$

$$\times \exp \left[\int_0^T \lambda(u) \, du - \int_{t_1}^T \lambda(u) \, du - \int_{t_2}^T \lambda(u) \, du ... - \int_{t_n}^T \lambda(u) \, du \right].$$

The characteristic functional $\Phi(\theta, T)$ corresponding to the interval $(0, T)$ can be evaluated using (2.6.6):

$$(2.6.7) \quad \Phi(\theta, T) = \left[1 - \int_0^T \lambda(u) \exp i\{\theta(u) - \int_u^T \lambda(v) \, dv\} \, du \right]^{-1} \{\exp - \int_0^T \lambda(u) \, du\}.$$

(iii) *General birth-and-death process*   We next consider the general birth-and-death process (see, for example, D.G. Kendall (1949)). The population state space is the positive real line, and the members of the population are characterized by their ages at any particular time t. At time $t = 0$, there is a primary of age x_0, and the primary, together with its secondaries (produced subsequently), generate a population in accordance with the following assumptions:

(1) the sub-populations generated by two coexisting individuals develop in complete independence of one another;
(2) an individual of age x present at time t has a probability $\lambda(x)\Delta + o(\Delta)$ of producing another individual of age zero in the interval $(t, t + \Delta)$, the probability of more than one birth in the interval $(t, t + \Delta)$ being $o(\Delta)$;
(3) an individual of age x present at time t has a probability $\mu(x)\Delta + o(\Delta)$ of death in the interval $(t, t + \Delta)$;
(4) the birth and death probabilities, $\lambda(.)$ and $\mu(.)$, respectively, do not depend on t, the time of existence.

For each t, $t > 0$, we have a point process generated by the members of the population (at t) being distributed over the interval $(0, x_0 + t)$. The birth-and-death process is Markovian in the sense that the probability distribution of individuals in ages at any time $t' > t$ depends only on the distribution at time t. For this reason the process is called a *Markov branching process* (see, for example, Harris (1963)). The characteristic functional of the process is given by

$$(2.6.8) \quad \Phi(\theta, t, x_0) = E\left\{\exp i \int_u \theta(u)\, d_u N(0, u; t)\right\},$$

where $N(0, u; t)$ is the random variable corresponding to the counting process and represents the number of individuals each of whose age lies in the interval $(0, u)$. We note that in view of our assumption (iii) (2), we have

$$(2.6.9) \quad \Pr\{N(y, h; t) \geqslant 1\} = 0 \quad \text{for} \quad y > x_0 + t, h > 0.$$

A differential equation for $\Phi(\theta, t)$ can be derived by analysing the course of events in the interval $(0, \Delta)$ of the parameter t. Note that the initial condition corresponding to an individual of age x_0 implies that

$$(2.6.10) \quad \Phi(\theta, 0, x_0) = e^{i\theta(x_0)}.$$

In the interval $(0, \Delta)$, there are three mutually exclusive possible outcomes:

(1) the individual survives up to Δ and gives birth to another in $(0, \Delta)$;
(2) the individual dies in $(0, \Delta)$;
(3) the individual merely survives up to Δ without producing any individual in $(0, \Delta)$.

Using the Markovian branching nature of the process, we find that

$$(2.6.11) \quad \Phi(\theta, t, x_0)$$
$$= (1 - \lambda(x_0)\Delta)(1 - \mu(x_0)\Delta)\Phi(\theta, t - \Delta, x_0 + \Delta) + \mu(x_0)\Delta$$

$$+ (1 - \mu(x_0)\Delta)\lambda(x_0)\Delta\Phi(\theta, t - \Delta, x_0 + \Delta)\Phi(\theta, t - \Delta, 0) + o(\Delta).$$

Proceeding to the limit as Δ tends to zero, we obtain

$$(2.6.12) \quad \frac{\partial\Phi(\theta, t, x_0)}{\partial t} = - (\mu(x_0) + \lambda(x_0))\Phi(\theta, t, x_0) + \mu(x_0)$$
$$+ \lambda(x_0)\Phi(\theta, t, x_0)\Phi(\theta, t, 0).$$

The above equation was derived by Bartlett and D.G. Kendall (1951). The general method of analysis as outlined above is due to Bellman and Harris (1948) and is known as the "regeneration point method". If we now take kth order variational derivatives of (2.6.12) as in section 2.4, we obtain the corresponding equations for the product densities. For instance, the first two derivatives yield

$$(2.6.13) \quad \frac{\partial h_1(x; t, x_0)}{\partial t} = - h_1(x; t, x_0)\mu(x_0) + \lambda(x_0)h_1(x; t, 0).$$

$$(2.6.14) \quad \frac{\partial h_2(x_1, x_2; t, x_0)}{\partial t}$$

$$= - \mu(x_0)h_2(x_1, x_2; t, x_0) + \lambda(x_0)h_2(x_1, x_2; t, 0)$$
$$+ \lambda(x_0)[h_1(x_1; t, x_0)h_1(x_2; t, 0) + h_1(x_1; t, 0)h_1(x_2; t, x_0)],$$

where

$$(2.6.15) \quad h_1(x; t, x_0) = E\{d_x N(0, x; t, x_0)\}/dx$$

$$(2.6.16) \quad h_2(x_1, x_2; t, x_0) = E\{d_{x_1} N(0, x_1; t, x_0)d_{x_2}(0, x_2; t, x_0)\}/dx_1 dx_2.$$

The explicit solution of the above equations can be obtained for special forms of $\lambda(.)$ and $\mu(.)$ (see Srinivasan and Koteswara Rao (1968)). The population problem was first posed in terms of these density functions by D.G. Kendall (1949) who derived integral equations for the covariance of $d_x N(0, x; t, x_0)$. A complete discussion of the problem can be found in Bartlett (1966) and Srinivasan (1969).

(iv) *Cascade processes* There is an important class of branching processes in which an individual (hereinafter called a *particle*) is transformed into two or more new particles which share some element of the original one (for example, mass or energy). There are a number of examples of this process in the realm of physics. The simplest one is the nucleon cascade that is observed in primary cosmic ray showers. The word "nucleon" is a common term in particle physics and is used to denote either a neutron or a proton. A highly energetic nucleon, on entering the top layer of the atmosphere from outer space, suffers a collision with the nuclei of the air and releases another nucleon from it. The secondary, whose energy was originally small enough to have it bound to the nucleus, acquires an appreciable portion of the primary energy. The primary and the secondary move further down the atmospheric layers and detach further nucleons. The problem is essentially one-dimensional, the parameter being t, the thickness of the atmospheric layer traversed by the nucleons (measured from above). At any particular value of t, we have a random number of nucleons distributed over the energy range $(0, E_0)$ where E_0 is the energy of the primary nucleon incident at $t = 0$. Thus we can introduce the counting process $N(0, E; t, E_0)$ corresponding to the number of nucleons present at the layer corresponding to thickness t, each having an energy lying in the range $(0, E)$. For a certain given set of assumptions governing the transfer of energy from the primary to the secondary and the average density of air nuclei, we can set up differential equations governing the characteristic functional or a set of such equations governing the product densities. For a detailed discussion, the reader is referred to the relevant chapters of Bartlett (1966) and Srinivasan (1969).

CHAPTER 2: PROBLEMS

2.1 Show that the relation (2.3.1) defines a one-to-one correspondence between $(\Phi, \mathbf{B}, P)$ and $(\mathfrak{N}, \mathbf{B}_N, P_N)$. (Moyal 1962)

2.2 Show that the relation (2.4.15) can be used to define the product-density generating functional $L[\xi]$ by

$$L[\xi] = 1 + \sum_{n=1}^{\infty} \lambda^n \int \xi(x_1)\xi(x_2) \dots \xi(x_n) h_n(x_1, x_2,\dots, x_n)\mu_n(dx^n).$$

How is $L[\xi]$ related to the characteristic functional?

(Kuznetsov and Stratonovich (1956))

2.3 Prove that $\Phi(\theta)$ as defined by (2.4.19) can be related to the moment distribution by

$$\Phi(\theta) = 1 + \sum_{n=1}^{\infty} \frac{i^n}{n!} \int \theta(x_1)\theta(x_2) \dots \theta(x_n) M_n(dx^n).$$

(Moyal (1962))

2.4 By considering the special case when the population is of size n and is distributed over A, show that (2.3.16) leads to

$$N^k = \Sigma \, C_l^k \, N(N-1) \dots (N-l+1).$$

Hence evaluate the constants C_l^k. (Ramakrishnan (1950))

2.5 Show that the mth moment of $N(t, x)$ as defined by (2.5.8) is given by

$$E\{[N(t, x)]^m\}$$

$$= \sum_{i_1=1}^{n} \sum_{i_2=1}^{n} \dots \sum_{i_s=1}^{n} \sum_{\substack{k_1, k_2,\dots,k_s=1 \\ k_1+k_2+\dots+k_s=m}}^{m} \sum_{l_1=1}^{k_1} \sum_{l_2=1}^{k_2} \dots \sum_{l_s=1}^{k_s} i_1^{k_1} i_2^{k_2} \dots i_s^{k_s}$$

$$\times C_{l_1}^{k_1} C_{l_2}^{k_2} \dots C_{l_s}^{k_s} \int_{t}^{t+x} \int_{t}^{t+x} \dots \int_{t}^{t+x} dx_1 \, dx_2 \dots dx_\nu \, h_{\nu[l_1, l_2,\dots, l_s]}(x_1, x_2, \dots, x_s)$$

$$(\nu = l_1 + l_2 + \dots + l_s),$$

where $h_{\nu[l_1, l_2,\dots,l_s]}(\dots.)$ represents the mixed product density of degree $\nu = l_1 + l_2 + \dots + l_s$ corresponding to there being l_α i_α-tuples $(\alpha = 1, 2,\dots, s)$.

(Srinivasan (1961))

2.6 *One-dimensional neutron model*   Neutrons can move forward and backward in a "rod" of length L and can be lost through the ends. Take X to be the interval $(0, L)$ and x the position of a neutron at "birth" (to be specified presently). A neutron born at x goes right or left with probability $\frac{1}{2}$ each. In any interval of length dx in the rod, it has probability $\alpha\, dx$ of being transformed into two neutrons, each having probability $\frac{1}{2}$ of going right or left independently of one another. Assuming that all neutrons move with constant velocity v and that at time $t = 0$ a neutron is injected into the left end (0), estimate the mean of the neutron density at any point x inside the rod at any time $t > 0$.

2.7 *Yule–Furry process* For a homogeneous Yule–Furry process, show that the product density $h_n(t_1,\, t_2,...,\, t_n)$ satisfies the equation

$$h_n(t_1,\, t_2,...,\, t_n)$$
$$= n\lambda h_{n-1}(t_1,\, t_2,...,\, t_{n-1}) + \lambda \int_0^{t_n} h_n(t_1,\, t_2,...,\, t_{n-1},\, \tau)d\tau$$

$$(t_1 < t_2 <... \, t_{n-1} < t_n).$$

Deduce that $h_n(t_1,\, t_2,...,\, t_n) = n!h_1(t_1)h_1(t_2) ... h_1(t_n)$ and hence derive the result corresponding to (2.6.7).   (Ramakrishnan and Srinivasan (1956))

2.8 The actual correlation functions $h_n(x_1,\, x_2,...,\, x_n)$ of a point process over reals are defined by

$$h_1(x_1) = h_1(x_1)$$
$$h_2(x_1,\, x_2) = h_1(x_1)h_1(x_2) + h_2(x_1,\, x_2)$$
$$h_3(x_1,\, x_2,\, x_3) = h_1(x_1)h_1(x_2)h_1(x_3) + 3\{h_1(x_1)h_2(x_2,\, x_3)\}_{\text{sym}}$$
$$+ h_3(x_1,\, x_2,\, x_3),$$

where $\{\ \}_{\text{sym}}$ are the symmetrical products of the quantities within the bracket. Show that $L[\xi]$ as defined in problem 2.2 is given by

$$L[\xi] = \exp\left[\sum_{m=1}^{\infty} \frac{1}{m!} \int \int \int ... \int h_m(x_1,\, x_2,...,\, x_m)\xi(x_1)\xi(x_2) ... \xi(x_m)\right.$$
$$\left. dx_1 dx_2 ... dx_m\right].$$

3 RENEWAL PROCESSES

3.1 Renewal theory

Renewal theory has its origin in the study of self-renewing aggregates and population growth, viewed mostly from a non-probabilistic angle. More recently the theory has centred on the study of general results relating to the sums of independent random variables. Feller (1941, 1948, 1949) made significant contributions to the theory and also gave the proper lead for the study of renewal processes. Smith (1957, 1958, 1960) further championed the methods of renewal theory, providing it with a firm analytical basis. A lucid and interesting account of the theory and its applications has been given by Cox (1962). The object of this chapter is to bring home the idea that a renewal process can be thought of as a special case of a stochastic point process, and that some of the results of renewal theory can be derived as special cases of theorems and results discussed in Chapter 2. In this section we shall summarize some of the main results relating to sums of random variables, in order to provide a proper perspective for the theorems and results relating to renewal processes. The reader who wishes to obtain a better insight into the subject is urged to study the work of Smith and the relevant chapters of Feller (1966).

Let $X_1, X_2, \ldots$ be a sequence of non-negative random variables which are independently and identically distributed. To avoid trivial cases, we shall assume that the X_i's do not vanish with probability one. To fix ideas, $\{X_i\}$ can be taken to represent the lifetime of machines that are being replaced. The first machine is installed at the instant $t = 0$ and is replaced instantaneously at time $t = X_1$. The replaced machine is again replaced at time $t = X_1 + X_2$ and so on. If we write $S_n = X_1 + X_2 + \ldots + X_n$, the partial sum S_n can be interpreted as the time at which the nth replacement is made. A reasonable object for investigation is the probability distribution of S_n. We can also use the symbol $N(t)$ to denote the largest number n for which $S_n \leqslant t$, so that $N(t)$ is the number of renewals that will have occurred up to and including the epoch t. One important result concerning $N(t)$ that has been well advertised is the *elementary renewal theorem* (see Feller (1941)) which states that the renewal function $H(.)$ defined by

$$(3.1.1) \qquad\qquad H(t) = E\{N(t)\}$$

has the property

(3.1.2) $\lim_{t \to \infty} H(t)/t = \mu_1^{-1}; \quad \mu_1 = E\{X_i\} \leqslant \infty,$

the limit being interpreted as zero if $\mu_1 = \infty$.

Blackwell (1948) generalized this result for sequences of random variables not having a lattice structure[*] and proved that

(3.1.3) $\lim_{t \to \infty} \{H(t + x) - H(t)\} = x/\mu_1$

for any fixed $x > 0$, the limit being again interpreted to be zero if $\mu_1 = \infty$. A further generalization was arrived at by Smith (1954); this is known as *Smith's key renewal theorem* and states that if $Q(t)$ is a non-increasing and non-negative function of positive t and Lebesgue integrable over $(0, \infty)$, then

(3.1.4) $\lim_{t \to \infty} \int_0^t Q(t - u)\, dH(u) = \dfrac{1}{\mu_1} \int_0^\infty Q(u)\, du.$

A number of useful conclusions about the characteristics of the renewal process can be drawn from the renewal theorems cited above, since they describe the asymptotic behaviour of $H(.)$. It will be shown that $H(.)$ (see section 3.2, below) adequately describes the characteristics of the renewal process. Thus the importance of these theorems can hardly be underestimated. From a mathematical point of view, two generalizations of renewal theory are possible by relaxing either of the following two restrictions:

   (i) The random variables $\{X_i\}$ are non-negative,
   (ii) The random variables $\{X_i\}$ are independently distributed.

If assumption (i) is relaxed and (ii) is retained, the process is commonly known as an *extended renewal process*. Blackwell's theorem remains true for an extended renewal process (Blackwell (1953)). Following Blackwell, it is convenient to write Z_1 for the first non-negative S_n and to define $Z_1 + Z_2 + ... + Z_m$ as being the first S_n not less than $Z_1 + Z_2 + ... + Z_{m-1}$. Feller (1968) has called the Z_i "ladder variables". The sequence $\{Z_n\}$ is a renewal process, and it is easy to deduce the results for the extended renewal process $\{X_n\}$ from the imbedded renewal process $\{Z_n\}$. Application of this idea yields rich dividends, and many of the results in the theory of queues and inventories can be deduced from the theorems and results of renewal theory (see, for example, Prabhu (1967)). There have been attempts to relax assumption (ii) and retain (i) (see, for example, Chow and Robbins (1963), Sankaranarayanan and Swayambulingam (1969)). At the outset it is useful to observe that in such a case we are dealing with a point process

---

[*] A sequence $\{X_i\}$ is said to possess a lattice structure if with probability one the greatest common divisor of $\{X_i\}$ is $w > 0$. In such a case the renewal process is said to be "discrete."

defined on the positive part of the real line, and some of the limit theorems may be visualized as the characteristics of a stationary point process. This aspect will be further discussed and the relevant results restated for stationary point processes in Chapter 4. Finally it is also useful to observe that the sequence of partial sums of $\{X_i\}$ form a sub-martingale, and hence the theorems and results of martingale theory can be interpreted appropriately.

3.2 Renewal processes

In the preceding section we considered a sequence of non-negative random variables $\{X_n\}$ which are independently and identically distributed. Very often it is useful to precede the sequence by another random variable X_0 which is also non-negative and independent of $\{X_i\}$, its distribution being not necessarily the same as that of any of the members of the sequence. The sequence $\{X_0, X_1, X_2, ...\}$ is usually called a *general renewal process,* or sometimes a *delayed renewal process* (see Feller (1968)). In terms of the replacement idea, X_0 can be interpreted as the residual lifetime of the machine in use at the time-origin $t = 0$. Alternatively we can define a general renewal process as a sequence of random points $t_1, t_2, ..., t_n,...$ on the positive real axis $(0, \infty)$, such that

(3.2.1) $$t_{i+1} - t_i = X_i \quad i = 1, 2,...,$$

$$t_1 = X_0,$$

where the successive intervals between the points are independently and identically distributed. Naturally X_0, which is the length measured from the origin (chosen arbitrarily) to the next random point, need not be distributed in the same manner as any of the $\{X_i\}$. The renewal function $H(t)$ introduced in the previous section is nothing but the expected number of random points in the interval $(0, t)$.

We can interpret the points as the objects (or individuals) of Chapter 2 and the population state as the collection of all the points on the half-line $(0, \infty)$. Thus we can seek a description of the point process in terms of the characteristic functional or the sequence of product densities. At this stage it is better to consider slight modifications of the problem by replacing the sequence $X_0, X_1, X_2,...$ by $\{X_i\}$, where i takes all integral values (both positive and negative), and considering a corresponding sequence of points on the real line $(-\infty, +\infty)$ (see Fig. 1). The random variable X_0 is still interpreted as being the residual lifetime of the machine in use at the origin (chosen arbitrarily). The point corresponding to the random variable X_{-1} can be interpreted as being the epoch corresponding to the most recent replacement considered from the same origin. Thus, barring X_0 and X_{-1}, all other members of the sequence $\{X_i\}$ are independently and identically distributed. Since the origin is an arbitrary point, it is better to reinforce the idea by marking

34

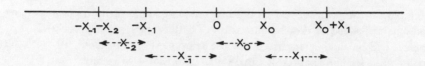

Fig. 3.1 The events corresponding to a renewal process plotted on the time axis

its coordinate as t with reference to a different origin (at $-t$ with reference to the original system). Thus, in the point process generated by the random variables $\{X_i\}$, we can consider, in the notation of Chapter 2, the counting process $N(t, x)$ representing the number of points in the interval $(t, t + x)$. Corresponding to the counting process, we can introduce the product densities $h_m(t, x_1, x_2,..., x_m)$ defined by

(3.2.2) $h_m(t, x_1, x_2,..., x_m)\, dx_1 dx_2 ... dx_m$

$$= \mathrm{E}\{d_{x_1} N(t, x_1) d_{x_2} N(t, x_2) ... d_{x_m} N(t, x_m)\}.$$

In particular, let us consider the product density of degree 2. This is defined by

(3.2.3) $h_2(t, x_1, x_2)$

$$= \lim_{\substack{\Delta_1 \to 0 \\ \Delta_2 \to 0}} \mathrm{E}\left\{\left[\frac{N(t, x_1 + \Delta_1) - N(t, x_1)}{\Delta_1}\right]\left[\frac{N(t, x_2 + \Delta_2) - N(t, x_2)}{\Delta_2}\right]\right\}$$

The right-hand side can be evaluated for $x_1 < x_2$ (or $x_1 > x_2$) if we make suitable assumptions about $F(.)$, the distribution function of $\{X_i\}$. If we wish to reproduce the results of conventional renewal theory, following Cox (1962, see p. 26), we shall assume that $F(0) = 0$, so that there are no concentrations of intervals of zero length with probability one. Thus we are confining our attention to regular point processes. In such a case it is clear that (3.2.3) can be written as

$$h_2(t, x_1, x_2) = \lim_{\substack{\Delta_1 \to 0 \\ \Delta_2 \to 0}} \mathrm{Pr}\,\{N(t, x_1 + \Delta_1) - N(t, x_1) = 1\}$$

$$\times \mathrm{Pr}\,\{N(t, x_2 + \Delta_2) - N(t, x_2) = 1 | N(t, x_1 + \Delta_1) - N(t, x_1) = 1\}/\Delta_1\Delta_2$$

(3.2.4) $= h_1(t, x_1)p(t, x_2, x_1) \quad (x_2 > x_1),$

where $p(t, x_2, x_1)dx_2\,(x_2 > x_1)$ represents the probability of occurrence of a random point in $(t + x_2, t + x_2 + dx_2)$ conditional upon the occurrence of a point at $t + x_1$. However, $p(t, x_2, x_1)$ can only be a function of $(x_2 - x_1)$ in view of the independence of the random variables $\{X_i\}\,(i \neq 0, -1)$ and hence we let $h(x_2 - x_1)$ denote that function. Now $h(.)$ has the following interpretation: $h(x)\Delta\,(x > 0)$ represents the probability that there is a random point in the interval $(x, x + \Delta)$, given that there is a random point at the origin. The function $h(.)$, defined for positive arguments, is called the *ordinary renewal density* (see Cox (1962)). The function $h_1(t, x)$ has a similar interpretation. The quantity $h_1(t, x)\Delta$ denotes the probability that there is a random point (not necessarily the first) in the interval $(t + x, t + x + \Delta)$ (t being arbitrary and $x > 0$). In other words, the quantity simply denotes the probability of occurrence of a random point (not necessarily the first) to the right of an arbitrarily chosen epoch t. The function $h_1(.,.)$ is called the *modified renewal density*.

Using exactly similar arguments, we find that the product density of degree m is given by

(3.2.5) $\quad h_m(t, x_1, x_2, \ldots x_m)$

$$= h_1(t, x_1)h(x_2 - x_1)h(x_3 - x_2)\ldots h(x_m - x_{m-1})$$

$$(x_1 < x_2 < \ldots < x_m).$$

Thus all the characteristics of the renewal process are determined if we are in explicit possession of $h_1(.,.)$ and $h(.)$. For instance, we can explicitly calculate the moments of the random variable $N(t, x)$ with the help of formula (2.3.16). Besides, the correlation of $N(t, x)$ over different intervals (overlapping or not) can also be calculated. For the explicit calculation, we need some of the basic properties of $h(.)$ and $h_1(t,.)$. The properties of these functions are best studied by means of the renewal equation. In the next section we shall derive the basic renewal equation and then proceed to discuss the solution in terms of the distribution of the basic random variables $\{X_i\}$.

3.3 The renewal equation and its ramifications

In our discussion thus far we have not explicitly made use of the probability density function governing the sequence of random variables $\{X_i\}$ through which the renewal process has been defined. The probability distribution of $\{X_i\}$ comes to the fore quite naturally if we attempt to spell out the characteristics of the two basic functions $h_1(t,.)$ and $h(.)$. We note that $h_1(t, x)\,dx$ denotes the probability that a random point (or event) occurs in the interval $(t + x, t + x + dx)$. Since this can arise in two mutually exclusive and exhaustive ways, according as it is the *first* random point or a subsequent random point, we note that $h_1(t,.)$ satisfies the equation

(3.3.1) $\quad h_1(t, x) = f_F(t, x) + \int\limits_0^x h_1(t, u)f(x - u)\,du,$

where

(3.3.2) $\quad f_F(t, x)\,dx = \Pr\{d_x N(t, x) = 1, \quad N(t, x) = 0\}.$

Thus $f_F(t, x)$ is the probability density governing the length of the interval from t to the next random point. In our notation the random variable representing the length of the interval may be taken to be X_0, provided we assume that the process has no past prior to t. However, there is an alternative way of characterizing the random variable. This tacitly assumes that the process has been in existence prior to t, in which case the random variable is known in the literature as the *forward recurrence time*. In a similar way the random variable X_{-1} to which we will advert later on is known as the *backward recurrence time*.

Next we observe that, using similar arguments, $h(.)$ satisfies the equation

$$(3.3.3) \quad h(x) = f(u) + \int_0^x f(u)h(x - u)\,du,$$

where $f(.)$ is the probability density function governing any of the variables $(\{X_i\}; i \neq 0, -1)$.

To make further progress, there are two courses of action open. The first consists in imposing a further condition on the process in the form of stationarity. The imposition of the condition is achieved by making t tend to infinity and assuming that $h_1(t,.)$ and $f_F(t,.)$ have limits which we shall denote by $h_1(.)$ and $f_F(.)$. Thus (3.3.1) can be written as

$$(3.3.4) \quad h_1(x) = f_F(x) + \int_0^x f(x - u)h_1(u)\,du.$$

We next note that we have an independent equation to determine $h(.)$, so that our problem is fully determinate if we are able to relate $f_F(.)$ to $f(.)$. To achieve this we note that equation (3.3.4) can be interpreted as follows. Since the first origin (i.e. the origin of the first choice, as specified by equation (3.3.1)), has receded to minus infinity, $h_1(x)\,dx$ denotes the probability that a random point occurs at a distance between x and $x + dx$ from any arbitrary chosen point (which is the new origin). $f_F(.)$ is interpreted similarly. Let the previous random point occur at a distance between u and $u + du$ measured in the negative direction. In view of stationarity, the probability of this event cannot depend on x and hence can only be $c\,dt$, where c is a constant. Thus we have

$$f_F(x) = c \int_0^\infty f(u + x)\,du$$

$$(3.3.5) \qquad\qquad = c(1 - F(x)),$$

where $F(.)$ is the distribution function corresponding to $f(.)$. Since we have assumed that the random variables $\{X_i\}$ take finite values with probability one, we have

$$(3.3.6) \qquad\qquad \int_0^\infty f_F(x)\,dx = 1.$$

The above equation identifies c as $1/\mu_1$, where

$$(3.3.7) \qquad \mu_1 = \int\limits_0^\infty x\, f(x)\, dx.$$

The alternative course, which is essentially due to Cox (1962), consists in obtaining an expression for $f_F(t, x)$, where t again has reference to an arbitrary origin, and then proceeding to the limit as t tends to infinity. Here again we analyse the course of events prior to the point t. The random point occurring in the interval $(t + x, t + x + dx)$ is either the first one counted from the origin or a subsequent one. If it is the first, then $t + x$ corresponds to the random variable X_0 whose probability density is $f_1(.)$. Thus we have

$$(3.3.8)^{(*)} \qquad f_F(t, x) = f_1(t + x) + \int\limits_0^t f(u + x)h_1(0, t - u)\, du.$$

The set of equations (3.3.1), (3.3.3) and (3.3.8) together constitute the description of *the general renewal point process which is not necessarily stationary*. If we proceed to the limit as t tends to infinity, we obtain

$$(3.3.9) \qquad \lim_{t \to \infty} f_F(t, x) = \frac{1 - F(x)}{\mu_1}.$$

It is convenient to choose $f_1(.)$ to coincide with the limit of $f_F(t,.)$. Cox calls such a process an "equilibrium renewal process". However, it should be borne in mind that any arbitrary choice of $f_1(.)$ leads to (3.3.9), provided that $f_1(t)$ tends to zero as t tends to infinity.

We note that the derivation of (3.3.5) or (3.3.9) involves some steps very similar to those used in the proof of the elementary renewal theorem as well as of the key renewal theorem. It is easy to conclude from the key renewal theorem that both $h_1(t)$ and $h(t)$ tend to the same limit $1/\mu_1$ as t tends to infinity. Later we shall see that this is a general property characteristic of any stationary point process (see section **4.3**). In view of this property we can apply the Laplace transform technique to express $h(.)$ and $h_1(.)$ in terms of $f(.)$. If $h^*(.)$, $h_1^*(.)$ and $f^*(.)$ are the corresponding Laplace transforms, we obtain from (3.3.3) and (3.3.4)

$$h^*(s) = \frac{f^*(s)}{1 - f^*(s)} \qquad \mathrm{Re}\, s > 0$$

$$f_F^*(s) = \frac{1 - f^*(s)}{s\mu_1}$$

$$(3.3.10) \qquad h_1^*(s) = (h^*(s) + 1)f_F^*(s) = 1/s\mu_1.$$

(*) The new function $h_1(0,.)$ can be re-expressed in terms of $f_1(.)$ and $h(.)$ (see problem 3.4, below).

Thus we are inevitably led to the conclusion that $h_1(.)$ is a constant equal to $1/\mu_1$. Here, again, Cox (1962) calls such a process an "equilibrium" renewal process". He has also visualized non-equilibrium processes for which the density function $f_F(.)$ is different from the choice specified by (3.3.9).

The distribution of the backward recurrence time can be discussed in a similar manner. In the general case, there are mutually exclusive types of events according to whether or not a random point occurs in the interval $(0, t)$. In the latter case, there is a discrete contribution to the density function in the form of a delta function. Denoting by $f_B(t,.)$ the probability density function of X_{-1}, we find that

$$(3.3.11) \quad f_B(t, x)$$
$$= \delta(t + x)(1 - F_1(t + x)) + \int_0^t h_1(0, u)(1 - F(t - u + x))\,du,$$

where $F_1(.)$ is the distribution function corresponding to the density function $f_1(.)$ and $\delta(.)$ is the Dirac delta function. If we proceed to the limit as t tends to infinity, we find that

$$(3.3.12) \quad \lim_{t \to \infty} f_B(t, x) = \lim_{t \to \infty} \int_0^\infty h_1(t - u)(1 - F(u + x))\,du$$
$$= \frac{1 - F(x)}{\mu_1}.$$

Thus we have proved that the forward and backward recurrence times are distributed identically in the case of a stationary renewal process. It does not follow that they are independently distributed. In fact there is a correlation between the two random variables (see problem 3.5, below).

3.4 The counting process and processes derived from it

We now proceed to complete the description of the counting process $N(t, x)$. A general formula will be deduced for the generating function by the techniques presented in section 2.3. The kth moment of $N(t, x)$ is given by (see formula (2.3.16))

$$(3.4.1) \quad E\,[N(t, x)]^k = E\,[\int_0^x d_u N(t, u)]^k$$
$$= \sum_{m=1}^k C_m^k\,m!\,\int_0^x h_1(t, x_1)\,dx_1 \int_{x_1}^x h(x_2 - x_1)\,dx_2$$
$$\cdots \int_{x_{m-1}}^x h(x_m - x_{m-1})\,dx_m,$$

where we have used the explicit expression (3.2.5) for the product density of degree m. As we have observed in an earlier section, the factorial moments of $N(t, x)$ are the convenient quantities to handle. Thus the kth factorial moment is given by

(3.4.2) $M_k(t, x) = \text{E}\,[N(t, x)]_k$

$$= k! \int_0^x h_1(t, x_1)\,dx_1 \int_{x_1}^x h(x_2 - x_1)\,dx_2 \ldots \int_{x_{k-1}}^x h(x_k - x_{k-1})\,dx_k.$$

The Laplace transform of $M_k(t, x)$ defined by

(3.4.3) $M_k^*(t, s) = \int_0^\infty M_k(t, x)\,e^{-sx}\,dx \quad \text{Re } s > 0$

can be expressed in terms of the Laplace transform of $h_1(t,.)$ and $h(.)$. By the key renewal theorem it follows that each of the functions $h_1(t, x)$ and $h(x)$ tends to the constant equal to $1/\mu_1$ as x tends to infinity. Thus the Laplace transforms $h_1^*(t, s)$ and $h^*(s)$ are defined in the domain Re $s > 0$. Since $M_k(t,.)$ is a convolution of $h_1(t,.)$ and a $(k-1)$-fold convolution of $h(.)$, we obtain

(3.4.4) $M_k^*(t, s) = k! \dfrac{h_1^*(t, s)}{s}\,[h^*(s)]^{k-1} \quad \text{Re } s > 0.$

Next we observe that if $G(u, t, x)$ is the probability generating function of $N(t, x)$, then we have

(3.4.5) $G(u, t, x) = 1 + \sum_{k=1}^\infty \dfrac{(u-1)^k}{k!}\,M_k(t, x) \quad |u - 1| < 1.$

It is easy to prove that there exists a positive number δ such that $G^*(u, t, s)$ is defined in the domain $0 < \text{Re } s < \delta$ and is given by

(3.4.6) $G^*(u, t, s) = \dfrac{1}{s} + \dfrac{h_1^*(t, s)(u-1)}{s[1 - (u-1)h^*(s)]}.$

On expressing $h^*(.)$ in terms of $f^*(.)$ (see equation (3.3.3)), we obtain

(3.4.7) $G^*(u, t, s) = \dfrac{1}{s} + \dfrac{h_1^*(t, s)[1 - f^*(s)](u-1)}{s[1 - uf^*(s)]},$

a formula obtained by Cox (1962).

From (3.4.1) we note that the second moment of $N(t, x)$ is given by

(3.4.8) $\text{E}\,[N(t, x)]^2 = \int_0^x h_1(t, u)\,du + 2 \int_0^x du \int_0^u h_1(t, u)h(v - u)\,dv.$

If we impose stationarity, we find that

(3.4.9) $\text{E}\,[N(0, x)]^2 = x/\mu_1 + (2/\mu_1) \int_0^x h(u)(x - u)\,du.$

The above result is true for a stationary point process, provided we interpret

$h(.)$ suitably. This point will be discussed in Chapter 4. A complete discussion regarding the calculation of moments and the probability distribution of $N(t, x)$ can be found in the monograph by Cox (1962).

Another interesting quantity characterizing the counting process is the correlation of the total number of renewals (counts) in different intervals. The correlation is best described by the generating function of the joint distribution of the counts corresponding to different intervals. We shall derive the generating function of the joint distribution of the counts corresponding to two non-overlapping intervals. Let $G_2(u, v, t, x, y)$ be defined by

$$(3.4.10) \quad G_2(u, v, t, x, y) = E\{u^{N(t,x)} v^{N(t+x,y)}\}.$$

Then the joint factorial moment of order (l, m) is given by

$$\mu_{l,m}(t, x, y) = \left(\frac{\partial}{\partial u}\right)^l \left(\frac{\partial}{\partial v}\right)^m G(u, v, t, x, y)\Big|_{u=v=1}$$

$$= \int_0^x dx_1 \int_0^x dx_2 \ldots \int_0^x dx_l \int_x^{x+y} dy_1 \int_x^{x+y} dy_2$$

$$(3.4.11) \qquad\qquad \ldots \int_x^{x+y} dy_m \, h_{l+m}(t, x_1, x_2, \ldots, x_l, y_1, y_2, \ldots, y_m),$$

where h_{l+m} is the product density of degree $(l+m)$ of random points on the segment (t, ∞). Expressing the product density in terms of the renewal density function as in (3.4.2), we note that

$$\mu_{l,m}(t, x, y) = l! \, m! \int_0^x dx_1 \int_{x_1}^x dx_2 \ldots \int_{x_{l-1}}^x dx_l \int_x^{x+y} dy_1 \int_{y_1}^{x+y} dy_2$$

$$\ldots \int_{y_{m-1}}^{x+y} dy_m \, h_1(t, x_1) \, h(x_2 - x_1)$$

$$(3.4.12) \qquad\qquad \ldots h(x_l - x_{l-1}) \, h(y_1 - x_l) \, h(y_2 - y_1) \ldots h(y_m - y_{m-1}).$$

Defining the Laplace transform of $\mu_{l,m}(t, x, y)$ as $\mu_{l,m}^*(t, s_1, s_2)$, we find, after some calculation,

$$(3.4.13) \quad \mu_{l,m}^*(t, s_1, s_2)$$

$$= l! \, m! \, \frac{[h^*(s_1)]^{l-1} \, [h^*(s_1) - h^*(s_2)] \, [h^*(s_2)]^{m-1}}{(s_1 - s_2)s_1} \, h_1^*(t, s_1).$$

As before, it is easy to show that $G_2^*(u, v, t, s_1, s_2)$ is defined in the domain $0 < \text{Re } s_1 < \delta$, $0 < \text{Re } s_2 < \delta$, $|u - 1| < 1$, $|v - 1| < 1$ by the relation

(3.4.14) $G_2^*(u, v, t, s_1, s_2)$

$$= \frac{1}{s_1 s_2} + \frac{(u-1)(v-1)[h^*(s_1) - h^*(s_2)]h_1^*(t, s_1)}{s_1(s_1 - s_2)[1 - (u-1)h^*(s_1)][1 - (v-1)h^*(s_2)]}.$$

It is easy to extend formula (3.4.14) to the general case.

Sometimes it is easy to calculate the correlation of the number of renewals directly without recourse to transform technique. For instance, the first-order correlation of the counts in the two intervals $(t, t + x)$ $(t + x, t + x + y)$ is given by

$$E[N(t, x)N(t + x, y)] = \int_{v=x}^{v=x+y} \int_{u=0}^{u=x} E[d_u N(t, u) \, d_v N(t, v)]$$

(3.4.15)

$$= \int_0^y du \int_x^{x+y} dv \, h_2(t, u, v).$$

On expressing $h_2(t,.,.)$ in terms of $h_1(t,.)$ and $h(.)$, we find that

(3.4.16) $E[N(t, x)N(t + x, y)] = \int_0^x h_1(t, u)[H(y + x - u) - H(x - u)]du,$

where $H(.)$ is the renewal function. If we let t tend to infinity, we note that $h_1(t,.)$ tends to a constant by the elementary renewal theorem and thus we obtain the following result, valid for a stationary renewal process:

(3.4.17) $\lim_{t \to \infty} E[N(t, x)N(t + x, y)]$

$$= \frac{1}{\mu_1}\left[\int_0^{x+y} H(u)du - \int_0^x H(u)du - \int_0^y H(u)du\right].$$

The above result is true also for a stationary point process, provided we interpret the function $h(.)$ appropriately. A formal proof will be provided in Chapter 4 when we discuss the properties of stationary point processes. We further observe that an extension of (3.4.17) to cover the case of general overlapping intervals is possible (see problem 3.8, below).

In practical applications, it is very often a process derived from the counting process rather than the counting process itself that plays an important ·role. For instance, in many physical problems, the response to the individual counts of the counting process is directly observable. A more practical physical quantity is the cumulative response. It is expedient to visualize the cumulative response over an interval $(t, t + T)$ as a linear functional of the counting process. A typical realization of the cumulative response corresponding to a given sequence of renewals in $(t, t + T)$ is given by

(3.4.18) $\phi_{[N]}(t, T) = \sum_i \psi(T - x_i),$

where $t + x_i$ denotes the coordinate of the ith renewal, and $\psi(.)$ is the response to the individual count of the renewal process and is defined to be zero for negative values of its argument. If we define the response by

(3.4.19) $\psi(x) = 1 \quad x > 0$

                              $= 0 \quad$ otherwise,

then the cumulative response is the counting process. The probability distribution of the cumulative response for general $\psi(.)$ has been the subject of investigation by physicists and electrical engineers from the turn of this century. Most of the results relate to the response of the Poisson counting process. (For an extensive review, see Rice (1944, 1945).) The above relation can be written as a stochastic Stieltjes integral,

$$(3.4.20) \quad \phi_{[N]}(t, T) = \int_0^T \psi(T - x) \, d_x N(t, x),$$

which expresses the linearity of the functional $\phi_{[N]}$. The first few moments of $\phi_{[N]}$ are easily evaluated in terms of the renewal densities $h_1(t, .)$ and $h(.)$:

$$(3.4.21) \quad E\left[\phi_{[N]}(t, T)\right] = \int_0^T \psi(T - x) h_1(t, x) dx$$

$$E\left[\phi_{[N]}(t, T)\right]^2$$

$$= 2 \int_0^T dx_1 \int_{x_1}^T \psi(T - x_1)\psi(T - x_2) h_1(t, x_1) h(x_2 - x_1) dx_1 dx_2$$

$$(3.4.22) \qquad + \int_0^T [\psi(T - x)]^2 h_1(t, x) dx.$$

For a stationary renewal process, the above expressions can be further simplified. With the help of the representation (3.4.18) it is also possible to calculate the autocorrelation of the cumulative response and hence the power spectrum.

We next observe that the response defined by (3.4.19) can be a random variable, and if we assume that the responses at different points are independently distributed, the moment formulae still hold good, provided we replace $\psi(T - x_1)$, $\psi(T - x_1)\psi(T - x_2)$ and $[\phi(T - x_1)]^2$ by their corresponding expectation values. There are practical situations where the individual responses are random functions of their arguments, these random functions associated with different points being not necessarily independent (see, for example, Srinivasan and Iyer (1966)). Some simple problems in which sums arise by the association of a random variable with each random point were considered by Doob (1948) in the context of renewal theory. Ramakrishnan (1953) considered sums of random variables associated with points of a general point process. In fact, formulae (3.4.19) and (3.4.20) are special cases of the general moment formulae derived by him. Later on,

Takacs (1956a) considered the general response phenomena arising from renewal processes and provided analytical proofs for the derivation of basic quantities of interest like the variance and the autocorrelation function of the cumulative response (see problems 3.12 and 3.13, below). Mercer and Smith (1959) considered cumulative processes arising from sum functions of random variables associated with a Poisson process. A brief account of the cumulative processes arising from renewal point processes is given in the monograph by Cox (1962).

3.5 Censors: application to the theory of counters

One of the important areas of applications of renewal processes is the description of type I and type II counters which relates to the study of the sequence of events selected from a stochastic point process. Let $\{X_i\}$ be a renewal process, and S_n be defined by

$$(3.5.1) \qquad S_n = X_1 + X_2 + \ldots + X_n.$$

We shall designate the events of the point process so generated as simply e-events. We introduce a selective filtering called a *paralytic censor* through an independent renewal process $\{Y_i\}$. Let n_1 be the least suffix n for which $S_n > Y_1$ and n_j the least suffix n for which $S_n > S_{n_{j-1}} + Y_j$. This defines a new sequence of events of a renewal process whose partial sums are defined by

$$(3.5.2) \qquad S_j^p = S_{n_j}.$$

We shall call the events of the filtered process or censored process "r-events" (r standing for recorded or response-yielding events) and denote the corresponding process by $\{r_{X_i}\}$. The selective filtering can be interpreted in terms of the familiar counter problem. At $t = 0$ an r-event occurs and a dead period extends to the epoch Y_1 up to which the counter does not register any of the e-events of the process $\{X_i\}$ that might have occurred. However, it registers the very next e-event subsequent to Y_1, and the e-event again gives rise to a dead period of duration Y_2, and so on. When the e-events form a simple Poisson process, we realize the type I counter.

There is another method of achieving selective filtering, known as the *guarantee censor*. In this case we define n_1 to be the least suffix for which $S_n > S_m + Y_{m+1}$ for all $0 \leqslant m < n$ (with $S_0 = 0$). In other words, each e-event generates a dead time, and only those e-events that are the left-hand endpoints of the resulting covered stretches of dead period are registered. When the e-events form a simple Poisson process, we realize the type II counter.

The literature on the subject is prolific and it is a difficult task to assign priorities. For a bibliography, we refer the reader to the article by Smith (1958), although there are a few omissions. It can, however, be unambiguously stated that it was Feller (1948) who recognized the relevance and usefulness of renewal theory to such problems. We shall derive explicit expressions for the renewal density of r-events or the probability density governing the duration of the interval between any two successive r-events.

First we consider the r-events generated by a paralytic censor (type I counter). Let $\psi(.)$ be the probability density function of Y_i. Then $f_r(.)$, the p.d.f. of intervals of the r-events, can be arrived at by classifying the "events" into two classes according as whether or not the next r-event is intercepted by an e-event. In the latter case the contribution to the density is $f_e(x) \times \int_0^x \psi(u)du$. In the former case we pick the last e-event in the series that occurs prior to the r-event. Imposing the condition that none of the e-events is an r-event, we find the contribution to the density from this class of events to be

$$\int_0^x h_e(u) f_e(t-u) \int_u^x \psi(v)dv,$$

so that we have

$$(3.5.3) \quad f_r(x) = f_e(x) \int_0^x \psi(u)du + \int_0^x h_e(u)f_e(x-u) \int_u^x \psi(v)dv,$$

where $h_e(.)$ is the ordinary renewal density of e-events. If we express $\psi(.)$ as a weighted average of negative exponential functions

$$(3.5.4) \qquad\qquad \psi(x) = \int_\lambda e^{-\lambda x} \lambda \phi(\lambda)d\lambda,$$

and if we write

$$(3.5.5) \quad \bar{h}_e^*(s) = \int_\lambda h_e^*(s+\lambda)\phi(\lambda)d\lambda, \quad h_e^*(s) = \int_0^\infty h_e(t)e^{-st} dt,$$

we obtain from (3.5.3)

$$(3.5.6) \quad f_r^*(s) = [h_e^*(s) - \bar{h}_e^*(s)]/[1 + h_e^*(s)]$$

or equivalently $h_r^*(.)$, the Laplace transform of the renewal density of r-events, is given by

$$(3.5.7) \quad h_r^*(s) = [1 + h_e^*(s)]/[1 + \bar{h}_e^*(s)] - 1.$$

If we specialize to the case when e-events form a simple Poisson process with parameter μ, (3.5.6) becomes

$$(3.5.8) \qquad\qquad f_r^*(s) = \mu \frac{\psi^*(s)}{s + \mu},$$

a result derived by Ramakrishnan (1954b). If, on the other hand, $\psi(.)$ is a negative exponential with mean $1/\lambda$, we find, from (3.5.7), that

$$(3.5.9) \quad h_r^*(s) = [1 + h_e^*(s)]/[1 + h_e^*(s+\lambda)] - 1.$$

The above result was derived by Smith (1957). All the relevant features of the r process can be brought out through (3.5.6) or (3.5.7).

We next consider the r-events generated by guarantee censor. In this case it is easy to express $h_r(.)$ directly in terms of $f_e(.)$ and $\psi(.)$. Let the r-event in $(x, x + dx)$ be intercepted by n e-events. Then, imposing the condition that each one of the dead periods of the e-events, as well as the r-event at the origin, does not extend up to x, we find that

$$h_r(x) = f_e(x)R(x) + \sum_{n=1}^{\infty} \int_0^x dt_1 \int_{t_1}^x dt_2 \ldots \int_{t_{n-1}}^x dt_n f_e(t_1) f_e(t_2 - t_1) \ldots$$

$$(3.5.10) \quad f_e(t_n - t_{n-1}) f_e(x - t_n) R(x) R(x - t_1) R(x - t_2) \ldots R(x - t_n),$$

where

$$(3.5.11) \qquad\qquad R(u) = \int_0^u \psi(v)\,dv.$$

If $f_e(.)$ is negative exponential with parameter μ, then we have

$$(3.5.12) \quad h_r(x) = \mu R(x) \exp\left[-\mu \int_0^x (1 - R(u))\,du\right],$$

a result due to Hammersley (1953) and Ramakrishnan (1954b). Hammersley (1953) considered a special problem in which the Poisson events are passed through a type II counter, and the filtered events are then passed through a type I counter. The final events recorded by the type I counter are the r-events. Takacs (1956b, c) discussed this problem in great analytical detail and proved the normality nature of the distribution of the number of r-events over a large period of observation. In addition he obtained an explicit expression for the probability distribution of the intervals formed by r-events when the primary process is Poisson.

To derive Takacs' formula, we note that (3.5.3) can be written in terms of the distribution function $F_r(.)$

$$(3.5.13) \quad F_r(x) = \int_0^x \psi(y)\,dy \int_y^x [1 - F_e(x - u)]h_e(u)\,du.$$

If we use the expression on the right-hand side of (3.5.12) for $h_r(.)$ and express $F_r(.)$ in terms of $h_r(.)$ through (3.5.8), we obtain the result derived by Takacs. On the other hand, if we use the right-hand side of (3.5.10) for $h_r(.)$, we obtain the general result when the primary process is not necessarily a Poisson process.

3.6 Regenerative processes

Palm (1943) introduced the idea of *gleichgewichtspunkt* (point of equilibrium) in a stochastic point process in the context of telephone traffic problems. This is closely related to the regeneration point introduced by Bellman and Harris (1948) and Janossy (1950) who dealt with population point processes. A special case of this was studied by Feller (1949) in his theory of recurrent events. Smith (1955) generalized Feller's results and was able to deal with more general stochastic processes possessing such regeneration points. These processes have come to be known as "regenerative processes". A formal, well-knit theory of such processes has been developed by Kingman (1964). In this section we shall give a short heuristic account of regeneration points and explain how the idea can be used to deal with certain continuous (parametric) stochastic processes.

Roughly speaking, a point of regeneration of a stochastic process $x(t)$ is an event $\Re$ characterized by the property that if it is known that $\Re$ happens at $t = t_1$, then the knowledge of the history of the process prior to t_1 loses its predictive value. In some special cases, the event $\Re$ is the only characteristic, so that the process regenerates itself with each occurrence of $\Re$. In more general cases, in addition to the occurrence of $\Re$, a knowledge of $x(t_1)$ is necessary for predictive purposes.

If we consider the renewal point process, it can be viewed as a general point process in which each event or incidence is a regeneration point. The occurrence of an event at $t = t_1$ uniquely determines the distribution of events in any collection of segments of points t such that $t > t_1$. If we further specialize to the case when the intervals between the successive events are exponentially distributed, we notice that any point (not necessarily an event) on the t-axis is a regeneration point. This result enables us to identify the stationary point process endowed with Markovian properties as the simple Poisson process.

To illustrate the role of the knowledge of $x(t_1)$ when the regenerative event $\Re$ occurs at $t = t_1$, let us consider a single-server queuing process. Customers arrive at points (in time) characterized by a renewal point process; the service times $\{y_i\}$ of the customers are assumed to be independently and identically distributed. Queue discipline is assumed to be strictly maintained. The total load $x(t)$ on the counter at any time is the main quantity of interest and is given by

$$x(t) = \min \left(x_0 + \sum_{i=1}^{n(t)} y_i - t, 0 \right),$$

where x_0 is the initial load at $t = 0$ and $n(t)$ is the number of customers who have joined the queue during $(0, t)$. Since we start with an arbitrary load (> 0) at $t = 0$ corresponding to one or more customers in the queue, including the one being served, the probability density function governing the time to the first arrival is also assumed to be specified along with the density function of the time-interval between any two successive arrivals. In this case the point corresponding to the first arrival, say at t_1, is a regeneration point in the sense that $\Re$ taken with $x(t_1)$ is sufficient for predictive purposes. This point will be illustrated in Chapter 8 where we present a detailed treatment of some of the queuing problems.

A special case of the above example was first presented by D.G. Kendall (1951) as a regenerative process. This corresponds to the case when the arrivals at the queue form a Poisson process. In such a case, the time-point at which $x(t) = 0$ is a regeneration point. In other words, any point in the time-interval during which the counter is idle (or free) is a regeneration point in the strict sense that the event $\Re$ is sufficient for predictive purposes. There are similar problems in storage theory that can be viewed as regenerative processes, and these will also be dealt with in Chapter 8. In branching phenomena, problems involving a knowledge of $x(t_1)$ were dealt with by Bartlett and Kendall (1951), and a full account of those problems can be found in the monograph by Srinivasan (1969).

3.7 Alternating renewal processes

Let us consider a random variable $Z(t)$, capable of assuming the values 0 and 1 so that the intervals of t in which $Z(t) = 0$ and $Z(t) = 1$ alternate regularly. Let these intervals in which $Z(t) = 0$ and 1 be independently distributed in such a manner that the intervals in which $Z(t) = i$ ($i = 0, 1$) have a common probability density function $f_i(.)$. Such a process, which can be called an "alternating renewal process", was first studied in detail by Takacs (1957b) (see also Cox (1962)). In fact, Kolmogorov (1949) considered a more general case in which $Z(t)$ takes values on an arbitrary set X, and he dealt with the sojourn time of the stochastic process in a subset $B \subset X$.

As a simple example of such a process, we can consider a machine subject to failure. As soon as the machine fails, the servicing unit takes it over for repair and reinstates it after some time. Thus we have an alternating sequence of time-intervals corresponding to running times and repair times of the machine. If the sequences of time-intervals are independent random variables, we have an alternating renewal process. Another example is provided by the paralysable (paralytic) counters discussed in section 3.5. If the particles arrive in the counter in a Poisson manner and if, on recording an arrival, the counter gets paralysed for a certain time (known as the "dead time"), we realize an alternating renewal process whose intervals correspond to the dead time on each recording and the time-interval from the end of the dead period until the next arrival. The latter type of intervals are of course exponentially distributed.

It is clear that the points at which $Z(t)$ has a jump ($1 \to 0$ or $0 \to 1$) generate a point process. We call these events of types a ($1 \to 0$) or b ($0 \to 1$). In the case of a machine subject to failure, the a-events correspond to the epochs of failure and the b-events to the epochs of reinstatement. By our assumption, it is clear that the marginal process consisting of a-events only (or b-events only) is a renewal point process whose intervals are characterized by the probability density function $f_1 * f_2(.)$. Thus the renewal densities of the marginal renewal process can be easily obtained.

Sometimes it may be important to know the state of $Z(t)$ (0 or 1) for any arbitrary t. It is useful to define $\pi_{ij}(t)$ by

(3.7.1) $\quad \pi_{ij}(t) = \lim_{\Delta \to 0} \Pr\{Z(t) = j \,|\, Z(0) = i, Z(-\Delta) \neq i\}, \quad i, j = 0, 1.$

All the relevant features of the alternating process can be described in terms of the stochastic process $Z(t)$. In the case of a machine subject to failure the main quantity of interest is the cumulative time spent in the state 1. For instance, the mean value of the time for which the machine has been in operation in $(0, t)$, given that at $t = 0$ the machine has just been started, is given b

(3.7.2) $\quad E\left[\int_0^t Z(u)\,du\right] = \int_0^t \pi_{11}(u)\,du.$

The π-functions satisfy the renewal equations

(3.7.3) $\quad \pi_{ij}(t) = \delta_{ij} \int_t^\infty f_i(u)\,du + \int_0^t f_i(u)\pi_{1-i,j}(t-u)\,du, \quad i, j = 0, 1.$

The above set of equations can be solved by using Laplace transform technique. In (3.7.1) it has been assumed that there is a transition of $Z(t)$ at $(-\Delta, 0)$. However, the problem can be tackled when the origin is an arbitrary point. For details the reader is referred to Cox (1962).

The alternating renewal process is characterized by the renewal densities $h_{ij}(.)$ defined by

(3.7.4) $\quad h_{ij}(t) = \lim_{\Delta, \Delta' \to 0} \Pr\{Z(t + \Delta) = j, Z(t) \neq j \,|\, Z(0) = i, Z(-\Delta') \neq i),$

$$i, j = 0, 1.$$

We accordingly have the renewal equations

(3.7.5) $\quad h_{ij}(t) = (1 - \delta_{ij})f_i(t) + \int_0^t f_i(u)\,h_{1-i,j}(t-u)\,du \quad i, j = 0, 1,$

which readily admit of a Laplace transform solution. In a similar manner we can also define modified and equilibrium renewal densities and derive the integral equations satisfied by them.

3.8 Markov renewal processes

In the previous section we restricted the state space of the random variable $Z(t)$ to be the discrete two-point space $(0, 1)$. If we enlarge the state space to n points, then we obtain a generalization of the alternating renewal process. As before, we can assume that the time-intervals in which $Z(t)$ continues to remain in different states are independently distributed such that

$$\lim_{\Delta \to 0} \Pr\{Z(t + x) = j, Z(t + u) = i : \forall u \leqslant x \,|\, Z(t) = i, Z(t - \Delta) \neq i\}$$

(3.8.1) $\qquad\qquad = f_{ij}(x) \quad i, j = 1, 2, \dots, n.$

If we literally interpret the transition of $Z(t)$ to be characterized by a change of state, then the quantities $f_{ii}(.)$ are zero functions. Relation (3.8.1) implies that the probabilities of transition obey the Markov property. The process can be thought of as a Markov chain for which the time-scale is transformed randomly. Such a process is called a "Markov renewal process", for obvious reasons. However, the implication that the $f_{ii}(.)$ are zero functions appears to be too restrictive, and a wider definition is possible in which the concept of $i \to i$ transitions is meaningful. We shall illustrate this point by an example from reliability analysis. Continuous production in a machine shop is maintained by keeping n redundant identical units. As soon as a machine fails, one of the n units on standby is switched on. Meanwhile the unit that has failed is taken out by the service unit for repair. As soon as the repair is over, the repaired machine is added to the pile of redundant units. It might happen that by the time the first unit has been repaired all the redundant units have failed (one after another). One of the important problems in reliability analysis will be the "down time" of the system, i.e. the time in which all the units are in process of being repaired. To fix ideas, let us assume that there is a single repair facility, the repair time for any individual unit being exponentially distributed, and that the items queue up for repair on a "first come, first attended" basis. Let the time to failure (measured from the epoch at which it is switched on) of any machine be distributed generally. In such a situation, the epochs corresponding to the failure of the machine are regeneration points. The history of the stochastic process prior to such a regeneration point $\Re$ can be completely ignored as regards its non-predictive content, provided we know the number of redundant units available at that epoch. We shall let $Z(t)$ denote the number of redundant units available. It is interesting to note that if at a particular regeneration point t_1, $Z(t_1) = m$, it is possible that $Z(t_2) = m$, where t_2 is the next regeneration point, although $Z(t)$ does not remain in the same state m in the time-interval (t_1, t_2). Thus the process is a Markov renewal process with $f_{ii}(.) \neq 0$.

It is better to define a Markov renewal process as a regenerative stochastic process $X(t)$ (see **3.6**) in which the epochs $\{t_i\}$ at which $X(t)$ visits any member of a certain countable set of states $\{A_\nu\}$ are regeneration points, the visits (consisting of the epoch and the state visited) being regenerative events. Thus at epoch t_m, the process $X(t)$ has visited the state A_i, and the next visit at epoch t_{m+1} is to the state A_j; the interval (t_m, t_{m+1}) characterizing the duration between successive visits to the members of the chosen set of states is governed by the probability density function $f_{ij}(.)$. With such a regenerative process, we can associate the stochastic point processes $N(t)$ generated by the epochs of visits of $X(t)$ to any one of the countable set of states. In a similar manner we can also associate with the process the point process $N_j(t)$ generated by the epochs of visits of $X(t)$ to the particular member j of the se of states. An interesting stochastic process $Z(t)$ can be associated with $X(t)$ a: follows. If t_m, t_{m+1}, t_{m+2} are three successive regeneration points, then

$$(3.8.2) \qquad Z(t) = j \quad t_m \leqslant t < t_{m+1}$$
$$= i \quad t_{m+1} \leqslant t < t_{m+2},$$

where A_j and A_i are respectively the states visited by $x(t)$ at $t = t_m$ and t_{m+1} The stochastic process $Z(t)$ is known as a "semi-Markov process", a terminology due to Smith (1955). Semi-Markov processes were introduced by Lévy (1954), Smith (1954) and Takacs (1954) independently. Smith (1954, 1955) established many of the important limiting properties of such processes on lines parallel to renewal theory. Pyke (1961a, b) studied in great detail the properties of the associated counting process when the states $\{A_i\}$ are countably finite.

If we confine our attention to Markov renewal processes in which the states are countably finite, then it is possible to study the point process $N(t)$ by the techniques presented in this chapter. For simplicity, let us denote the state A_j by the index j. If all the states are recurrent, then $N(t)$ can be thought of as a stationary point process by shifting the time origin to minus infinity. The analogy with the renewal process is complete in the sense that the point process can be described by the renewal density $h_{ij}(.)$, where

$$(3.8.3) \quad h_{ij}(x) = \lim_{\Delta \to 0} E \{d_x N_j(0, x) | N_i(-\Delta, \Delta) = 1\}/dx;$$

here we use the usual notation $N_i(0, x)$ to denote the number of events (whic are regenerative) of type i in the interval $(0, x)$. Thus the renewal density can be thought of as a matrix in the state space. Let us use the notation $f_{ij}(.)$ to denote the corresponding probability density functions of the time-intervals

$$(3.8.4) \quad f_{ij}(x)$$

$$= \lim_{\Delta \to 0} \Pr \{d_x N_j(0, x) = 1, d_u N_j(0, u) = 0 \,\forall\, u \in (0, x) | N_i(-\Delta, \Delta) = 1\}/dx.$$

Then it is easy to see that $h_{ij}(.)$ satisfies the equation

$$(3.8.5) \quad h_{ij}(x) = f_{ij}(x) + \sum_{k=1}^{n} \int_0^x f_{ik}(u) h_{kj}(x - u) du, \quad i, j = 1, 2, \ldots, n,$$

which is the matrix analogue of the celebrated renewal equation introduced in **3.3**. Thus all the results of renewal theory are capable of extension.

It is also useful to study the random variable $Z(t)$ defined by (3.8.2) by defining $\pi_{ij}(t)$ as

$$(3.8.6) \quad \pi_{ij}(t) = \lim_{\Delta \to 0} \Pr \{Z(t) = j \mid Z(0) = i, Z(-\Delta) \neq i\}.$$

With the help of $\pi_{..}(.)$, it is possible to describe the properties of sojourn time in different states. The regenerative nature of the epochs of entry into the states enables us to obtain

$$(3.8.7) \quad \pi_{ij}(t)$$

$$= \delta_{ij} \sum_{k=1}^{n} \int_{t}^{\infty} f_{ik}(u) \, du + \sum_{k=1}^{n} \int_{0}^{t} f_{ik}(u) \pi_{kj}(t-u) \, du, \quad i, j = 1, 2, \ldots, n.$$

The special case when $n = 2$ has attracted a good deal of attention (see for example Takacs (1957a, b), Cox and Lewis (1966)).

CHAPTER 3: PROBLEMS

3.1   Use the law of large numbers to prove (3.1.2).

3.2   Show that $H(t) - \dfrac{t}{\mu_1} \to \dfrac{\mu_2}{2\mu_1^2} - 1$ as $t \to \infty$, where μ_2 is the second moment of $f(.)$. (Smith, 1954)

3.3   Prove that stationarity of the process implies that the random variables X_0 and X_{-1} introduced in **3.3** are identically distributed. (Cox, 1962)

3.4   Show that the function $h_1(.,.)$ introduced in **3.3** satisfies the equation

$$h_1(0, x) = f_1(x) + \int_0^x f_1(u)\, h(x - u)\, du.$$

3.5   For a stationary renewal process, obtain an explicit expression for the joint probability density of X_0 and X_{-1} and show that

$$\text{Cov}\,[X_0, X_{-1}] = \frac{\mu_3}{6\mu_1} - \frac{1}{4}\left(\frac{\mu_2}{\mu_1}\right)^2,$$

where μ_2 and μ_3 are respectively the second and third moments of the interval distribution (of the function $f(.)$). Deduce a sufficient condition for the perfect prediction of X_{-1} given X_0.

3.6   Show that for an ordinary renewal process equation, (3.4.7) takes the form

$$G^*_{\text{ord}}(u, s) = \frac{1 - f^*(s)}{s(1 - u f^*(s))},$$

while for an equilibrium renewal process G^* is given by

$$G^*_{\text{eqm}}(u, s) = \frac{1}{s} + \frac{u - 1}{\mu_1 s} G^*_{\text{ord}}(u, s). \tag{Cox, 1962}$$

3.7   Extend formula (3.4.8) corresponding to the counts in two overlapping intervals.

3.8   Obtain an extension of formula (3.4.17) by evaluating

$$\lim_{t \to \infty} \text{E}\,[N(t, x)N(t, y)].$$

3.9   If $\pi(n, t) = \lim\limits_{\Delta \to 0} \text{Pr}\,\{N(0, t) = n \mid N(-\Delta, 0) = 1\}$, where $N(0, t)$ is the random variable representing the number of renewals in $(0, t)$, show that

$$\pi(n, t) = -\int_0^t \pi(n - 1, u)\phi'(t - u)\,du + \delta_{n0}\phi(t),$$

where $\phi(t)$ is the survivor function defined by

$$\phi(t) = \int_t^\infty f(u)\,du.$$

Hence deduce that

$$G_{\text{ord}}^*(u, s) = [s\{1 - (u - 1)h^*(s)\}]^{-1}$$

(Ramakrishnan and Mathews, 1953)

3.10   Show that for a counter of type II with a constant dead time a and a primary Poisson process with unit rate,

$$G(u, t) = 1 + \frac{(u - 1)\,e^{-a}(t - a)}{1!} + \ldots + \frac{(u - 1)^N\,e^{-Na}(t - Na)^N}{N!},$$

where $Na < t < (N + 1)a$.   (Ramakrishnan, 1951)

3.11   Imagine a general counter in which the registered and unregistered events are followed by two different constant dead times, respectively equal to a and b. If a Poisson stream of events with unit rate ($\lambda = 1$) is incident on it, show that the Laplace transform of the ordinary renewal density $h(.)$ is given by

$$h^*(s) = [s\,e^{as+b} + e^{(a-b)s} - 1]^{-1}.$$

Hence derive explicit expressions for $\pi(n, t)$ and the expected value of the number of recorded events in $(0, t)$.   (Ramakrishnan and Mathews, 1953)

3.12   Consider the stochastic process

$$\eta(t) = \sum_{n=1}^\infty g(t - S_n, \chi_n),$$

where $g(t - S_n, \chi_n)$ is the response of a physical system at time t due to an impulse at time S_n, the response depending on a random parameter χ_n. Let $g_1(u) = E^* g(u, \chi_n)$ and $g_2(u, v) = E^* g(u, \chi_n)g(v, \chi_n)$, where E^* is the expectation symbol with respect to χ_n. If S_n is the partial sum of the renewal process X_n, and the χ_n's are independently distributed, show that

$$\lim_{t \to \infty} E\,[\eta(t)] = \mu_1^{-1} \int_0^\infty g_1(u)\,du,$$

$$\lim_{t \to \infty} \text{Cov}\,[\eta(t), \eta(t + \tau)]$$

$$= \mu_1^{-1} \int_0^\infty g_2(u, u + \tau)\,du + \mu_1^{-1} \int_0^\infty \int_0^u g_1(u)g_1(v)\psi_\tau(u - v)\,du\,dv,$$

where

$$\psi_\tau(u - v) = h(|u - v + \tau|) + h(|u - v - \tau|) - 2\mu_1^{-1}.$$

(Takacs, 1956a; Smith, 1958)

3.13   In problem 3.12, choose $g(u, x) = xe^{-\alpha u}$ $(\alpha > 0)$ and show that the autocorrelation function of the process is proportional to

$$(\mu^2 + \sigma^2)e^{-\alpha|\tau|} + 2 \int\limits_{-\infty}^{+\infty} e^{-\alpha|t-\tau|} k(t) dt$$

where $k(t) = h(|t|) - \mu_1^{-1}$ and $\mu = E[\chi_n]$.   (Takacs, 1956a; Smith, 1958)

4   STATIONARY POINT PROCESSES

4.1   General definitions

In the previous chapter we considered a special class of point processes generated by sums of non-negative independent variables. It is interesting to investigate to what extent the properties discussed in Chapter 3 remain valid if the statistical independence of the random variables is relaxed. In such a case we are dealing with a point process defined on the real line. The process then becomes too general, and nothing tangible is achieved by translating the general formulae derived in Chapter 2 corresponding to the special case $X = R$. To make further progress we shall require the process to be *stationary*. Of course this does not mean that only stationary processes are endowed with many special properties. In fact there are other types of non-stationary processes like branching processes, which can be studied under the general class of evolutionary point processes. A comprehensive account of such processes with special reference to their regenerative nature can be found in the monographs of Harris (1963) and Srinivasan (1969). In this chapter we shall attempt to characterize point processes that exhibit some kind of stationary behaviour. Besides these there are also transient point processes whose behaviour eventually becomes stationary (time-invariant), and we will have occasion to study some of the physically significant class of such processes in Chapter 5.

The properties of stationary point processes were first studied by Wold (1948a) and Bartlett (1954) to whom we owe the current terminology. Earlier, Rice (1945) had studied the point process generated by the zero crossings of a Gaussian process, and had obtained a relation between the product densities of the zero crossings and the distribution of the interval between any two successive crossings. Khintchine (1955), in his attempt to systematize the methods of queuing theory, provided an elegant approach to stationary streams of uniform events. Later, McFadden (1962) continued the work and derived many of the general properties of stationary point processes via the distribution functions of Palm (1943) and Khintchine (1955). About the same time, Kuznetsov and Stratonovich (1956) studied problems relating to the theory of correlated random points by using certain correlation functions which are nothing but the product densities of Ramakrishnan (1950). We now propose to give a detailed and systematic account of these developments in the theory of point processes in the light of the general methods developed in Chapter 2.

To start with, we note that there are a number of types of stationarity which need to be defined carefully. Using the notation of Chapters 2 and 3, let $N(t, x)$ be the random variable representing the number of points or event (hereinafter we shall use the word "event") in the intervals $(t, t + x)$ of the counting process. We say that the point process is *simply stationary* if

(4.1.1) $\Pr \{N(t, x) = n\}$

$$= \Pr \{N(t + h, x) = n\} \quad \text{for all } t, x, \quad h > 0 \text{ and } n \geqslant 0.$$

Thus, simple stationarity implies invariance of the probability distribution of the events in a fixed interval under arbitrary translation.

There is another type of stationarity known as *weak stationarity*. A point process is said to be weakly stationary if

(4.1.2) $\Pr \{N(t_1, x_1) = n_1, N(t_2, x_2) = n_2\}$

$$= \Pr \{N(t_1 + h, x_1) = n_1, N(t_2 + h, x_2) = n_2\}$$

$$\text{for all } t_1, t_2, x_1, x_2, \quad h > 0, n_1 \geqslant 0, n_2 \geqslant 0.$$

In other words, the joint distribution of events (of a weakly stationary process) in two fixed intervals remains invariant under arbitrary translation.

The third type of stationarity, known as *complete stationarity*, holds for a point process if

(4.1.3) $\Pr \{N(t_i, x_i) = n_i, \quad i = 1, 2, \dots m\}$

$$= \Pr \{N(t_i + h, x_i) = n_i, \quad i = 1, 2, \dots m\}$$

for all $t_i, x_i, n_i \geqslant 0, (i = 1, 2, \dots, m), h > 0$, and for all positive integral values of m.

In our definitions of stationarity no mention has been made of the location of the time origin. This is because there are two possibilities for a stationary point process.

(i) The process is started at $t = 0$ with appropriate initial conditions that produce stationarity. Such initial conditions are known as *stationary initial conditions*. Let us amplify the point with respect to renewal processes. Consider the renewal process defined by the sequence of independent random variables $\{X_i\}$, where X_0 is distributed differently from the other members of the sequence, these being distributed identically with common probability density function $f(.)$. Let the probability density function of X_0 be $f_1(.)$. If we choose $f_1(.)$ as $[1 - F(.)]/\mu_1$, the process becomes a stationary renewal process.

(ii) The process is transient and is visualized beyond $t = 0$ as the starting point moves off to the left. As the starting point goes to minus infinity, the process becomes stationary. The specification of the state of the process at

$t = 0$ is known as the *stationary equilibrium condition*. A typical example is the equilibrium renewal process defined in section **3.3**, where the process was constructed with reference to an arbitrary origin and then the limit as t tends to infinity was obtained. This is equivalent to shifting the origin to minus infinity.

There is also a third possibility in which we completely circumvent the specification of the origin by accepting the stationary point process as such. In such a case the point $t = 0$ is an arbitrary point which need not synchronize with the occurrence of the event. Such a method of choosing $t = 0$ is known as *asynchronous sampling* of the process, a terminology due to Cox and Lewis (1970).

To discuss the stream of events studied by Wold (1948a, b), Khintchine (1955) and McFadden (1962), we consider a discrete parameter stochastic process $\{y_n\}$ ($n = 0, \pm 1, \pm 2, ...$) associated with the probability space $(\Phi, \mathbf{B}, P)$. We would of course require that y_n be finite valued and non-negative with probability one. Let x_n be defined by

$$(4.1.4) \qquad x_n = \sum_{0}^{n} y_k \qquad n \geqslant 0$$

$$= y_0 - \sum_{n}^{-1} y_k \qquad n \leqslant -1.$$

Then it is easy to see that the σ-subfields induced on $(\Phi, \mathbf{B}, P)$ by $\{x_n\}$ and $\{y_n\}$ are identical. Thus, in the notation of section **2.1** (where X is restricted to be the set of reals), the point process can be described directly in terms of randomly located points. The quantities of interest are the following random variables:

    (i) $N(t, x)$, the number of points that fall in the interval $(t, t + x)$;
    (ii) $L_n(t)$, the time required for the nth point after t to occur;
    (iii) $L_{-n}(t)$, the time required for the nth point prior to t to occur.

The first random variable characterizes the process essentially as a counting process. The second, for varying n, identifies the process in terms of the lengths of successive intervals. The last variable brings to the fore some kind of a prehistory in terms of successive intervals again. Since each of the random variables provides an adequate description of the process, the notion of stationarity can be introduced through the invariance of the distribution functions governing each of the random variables under arbitrary translation of the one-dimensional parameter. This has been done by Beutler and Leneman (1966) who have shown that the invariance of the distribution function of any one of three random variables under arbitrary translation of the parametric values implies the invariance of the distribution function of each of the other two variables.

In what follows we shall assume that the point process is completely stationary. We shall use the following notation:

(4.1.5) $$g_n(x)\,dx = \Pr\{x \leqslant L_n(t) \leqslant x + dx\}$$

(4.1.6) $$f_n(x)\,dx = \Pr\{x \leqslant S_n \leqslant x + dx\},$$

where
$$S_n = \sum_{i=k}^{n+k-1} (x_{i+1} - x_i)$$

$$= \sum_{i=k}^{n+k-1} y_i \quad (k \text{ an arbitrary integer})$$

(4.1.7) $$p(n, x) = \Pr\{N(t, x) = n\}.$$

We shall also use the symbols $G_n(.)$ and $F_n(.)$ to denote the distribution functions corresponding to the density functions $g_n(.)$ and $f_n(.)$ respectively. The functional symbols $f(.)$ and $F(.)$ will be used to denote $f_1(.)$ and $F_1(.)$.

4.2 Distribution functions and general identities

Palm (1943) introduced the function corresponding to the probability of absence of events in $(t, t + x)$ conditional upon the occurrence of an event at time t. In our notation this probability is given by $1 - F(x)$. Khintchine (195? extended this idea and introduced a sequence of density functions defining the probability that k events occur in a given interval, conditional upon an event having occurred at the beginning of the interval. In our notation, the $g_{-n}(.)$ are the Palm–Khintchine functions.

We shall first investigate the consequences of stationarity on the distribution of $L_n(t)$ and S_n. To arrive at the distribution of $L_n(t)$, we choose $t = 0$ corresponding to the asynchronous sampling of the process. Because of station arity, the event prior to the origin occurs at a distance (measured backwards) between u and $u + du$ with probability $C\,du$, C being a constant. Thus we have, for $n \geqslant 1$,

$$g_n(x)\,dx = \int_{y=0}^{x} \Pr\{y < L_1 < y + dy, \, x - y < S_{n-1} < x - y + dx\}$$

$$= C\,dx \int_{0}^{\infty} \Pr\{x + u < S_n < x + u + du, \, S_{n-1} \leqslant x\}$$

(4.2.1) $$= C\,dx\,[F_{n-1}(x) - F_n(x)].$$

Since the constant C is independent of n, we determine C by taking $n = 1$ and using the condition

(4.2.2) $$\int_{0}^{\infty} g_n(x)\,dx = 1$$

which expresses the fact that the random variables y_i are finite valued with probability one. Thus we have

(4.2.3) $$1/C = \int\limits_0^\infty [1 - F(x)]\,dx = \mathrm{E}\{y_i\} = \mu_1.$$

To obtain the density function of L_{-n}, we note that in view of complete stationarity we have

(4.2.4) $$\Pr\{x < L_{-n} < x + dx\} = C \int\limits_0^x dx \,\Pr\{y < S_{n-1} < y + dy, S_n > x\}$$
$$= Cdx\,[\int\limits_0^x f_{n-1}(y)\,dy - F_n(x)]$$
$$= C\,[F_{n-1}(x) - F_n(x)]\,dx,$$

so that we have

(4.2.5) $$g_{-n}(x) = C\,[F_{n-1}(x) - F_n(x)],$$

where the constant can again be identified with $1/\mu_1$.

We have thus proved that $L_n(t)$ and $L_{-n}(t)$ are identically distributed, provided the process is completely stationary. If we put $n = 1$, we obtain

(4.2.6) $$g_{-1}(x) = g_1(x) = [1 - F(x)]/\mu_1,$$

a result which generalizes that of renewal theory (see equations (3.3.1) and (3.3.9)).

Next we establish some identities connecting the probabilities $p(.,x)$ and the Palm–Khintchine functions $G_.(x)$. By definition,

$$p(0,x) = \Pr\{L_1 > x\}$$

(4.2.7) $$= 1 - G_1(x).$$

More generally we have

$$p(n, x) = \Pr\{L_{n+1} > x, L_n < x\}$$
$$= \Pr\{L_{n+1} > x\} - \{\Pr L_n > x\}$$

(4.2.8) $$= G_n(x) - G_{n+1}(x).$$

The above relation was first derived by Khintchine (1955) in a different notation. Ramakrishnan (1954c) deduced such relations for stationary renewal processes in connection with one-dimensional molecular distribution.

To end our discussion on distribution functions, we note that $G_.(x)$ are expressible in terms of $F_.(x)$ and that $p(.,x)$ is again expressible in terms of $G_.(x)$ and hence of $F_.(x)$. Thus the description of the process in terms of any one of the random variables $N(t, x)$, S_n and $L_n(t)$ is completely equivalent to that in terms of any other. We further note that the distribution functions $F_n(.)$ or $G_n(.)$ and $p(n,.)$ also bring to the fore the correlation of the random variables $\{y_i\}$ corresponding to the intervals formed by different successive events (or points) (see problem 4.4, below). However, the correlation or dependence of the random variables is reflected in the univariate distribution functions only to a limited extent. In order to characterize the dependence completely, it is necessary to study the properties of stationary product densities of different orders. This we shall presently do.

4.3 Dependence and correlation: generalized renewal and product densities

We now proceed to discuss the correlational structure of stationary point processes. There is a regularity property associated with a certain class of processes. Crudely speaking, such a property is characterized by the non-occurrence of multiple events or clusters of points. Khintchine called such a process "an orderly stream of events" since the events can be ordered with probability one; Cox and Lewis (1970) designate this property as "regularity". A process is said to be "regular" or "orderly" if it satisfies the condition

$$(4.3.1) \qquad \frac{\Pr\{N(t, \Delta) = n\}}{\Pr\{N(t, \Delta) = 1\}} = O(\Delta), \quad n > 1.$$

Orderliness can also be characterized in terms of the probability distribution of the random variables $\{y_i\}$ or $\{S_n\}$, introduced in section 4.1 (see problem 4.5, below).

To get an orientation, let us consider a simple stationary Poisson process (Khintchine calls such a process "a simple stream of uniform events"). Such a process can be described as either

    (i)   a process in which the number of events in a period of length t is distributed according to the Poisson law with a parameter proportional to t, or

    (ii) a process without after-effect (memory) such that L_1 as defined in **4.1** is exponentially distributed.

If we use specification (i) for a general stationary process, it may be necessary to obtain the multivariate distribution of the random variables $N(t_i, x_i)$. Of course the best way to characterize the process is through the characteristic functional, but this may not be possible in all cases.

If we resort to specification (ii), there are two alternatives. Since the presence of after-effects is an important correlational property, a process can be specified by a complete prehistory up to the instant t under consideration. Varying degrees of non-Markovian nature may reduce the degree of indispensability of the complete prehistory. For instance, if the process is a renewal process, the prehistory need extend only up to the previous event. For this reason Khintchine calls a renewal process "a process with limited after-effect". For a non-renewal process it may be necessary to specify the history even beyond the previous event. This point will be illustrated by an example in section **6.4**. With the help of the prehistory, it is possible to derive the probability of occurrence of an event in the immediate future. Thus the process can be characterized by the *intensity function*. It will turn out that this is nothing but the product density function, suitably interpreted for a general non-regular process.

The other alternative is through the probability structure of successive intervals between events. This approach has been used by many workers in different fields, since it is perhaps the most direct (see, for example, McFadden (1962), McFadden and Weissblum (1963)). The intensity functions are characterized by means of the structure of interval distributions. In fact this is done even in the case of branching phenomena (see, for example, Messel (1954)). Since the different methods of description are essentially equivalent, we shall find it advantageous to use the characterization in terms of intensity functions.

The *complete intensity function* at t corresponding to the complete history $\mathcal{H}_t$ of the past $(-\infty, t)$, which we shall denote by $\lambda(t|\mathcal{H}_t)$, is defined by

$$\lambda(t|\mathcal{H}_t) = \lim_{\Delta \to 0} \Pr\{N(0, t + \Delta) - N(0, t) \geqslant 1 | \mathcal{H}_t\}/\Delta$$

(4.3.2) $$= \Pr\{d_t N(0, t) \geqslant 1 | \mathcal{H}_t\}/dt.$$

As remarked earlier, in any particular problem it is important to determine the minimum prehistory up to t, the epoch under consideration, rather than the complete history, which may not be necessary to determine the probability structure in the immediate future after t. Since this minimum will vary from problem to problem, it is necessary to explore the possibility of general characterization of the probabilistic structure in terms of only a partial history. A complete prehistory will amount to giving the location of the sequence of events prior to t. Rather, we can use limited information by giving the location of some events only, without reference to other events in the interval $(0, t)$, the origin being chosen arbitrarily and our interest centring only on the events occurring subsequent to the point (epoch) corresponding to the origin. Such a description is best provided in terms of the sequence of product densities $h_1(t_1), h_2(t_1, t_2), \ldots$ (conditional upon a certain prehistory), where $t_1, t_2, \ldots$ are points in the interval $(0, t)$. Accordingly we can attempt to provide a description of the process in terms of the product densities and this we proceed to do.

We first note that stationarity implies that $h_1(u)$, the generalized product density of degree one of events, is a constant. We have already made use of this result in determining the probability density function governing L_n or L_{-n}. For notational convenience we shall denote the constant by μ, so that

(4.3.3) $$h_1(u) = \mu = 1/\mu_1.$$

The mean number of events in an arbitrary interval $(0, t)$ is given by

$$(4.3.4) \qquad E[N(0,t)] = \int_0^t h_1(v)\,dv = \mu t,$$

which is similar to the result in renewal theory. Thus μ is the intensity of the stream of events. If we specify the various stationary probabilities for multiplicities by

$$(4.3.5) \qquad \lim_{\Delta \to 0} \Pr\{N(u, \Delta) \geqslant 1\}/\Delta = \lambda,$$

$$(4.3.6) \qquad \lim_{\Delta \to 0} \Pr\{N(u, \Delta) = i\}/\Delta = \lambda_i,$$

so that

$$(4.3.7) \qquad \sum_{i \geqslant 1} \lambda_i = \lambda,$$

we notice that

$$(4.3.8) \qquad \lambda \leqslant \sum_i i\lambda_i = \mu.$$

Thus we can conclude that $\lambda = \mu$ if and only if

$$(4.3.9) \qquad \Pr\{N(u, \Delta) = i\} = o(\Delta) \quad i > 1.$$

Condition (4.3.9), however, characterizes the regularity property. Thus we have proved that *a stationary point process is regular if and only if $\mu = \lambda$.* This result is known as *Korolyuk's theorem* (see Khintchine (1955)).

We can derive from (4.3.3) another interesting result. By the definition of $h_1(u)$, we have

$$(4.3.10) \qquad h_1(u) = \sum_1^\infty g_n(u).$$

Applying the stationarity condition and shifting the origin to $-\infty$, we note that

$$(4.3.11) \qquad \lim_{u \to \infty} \sum_1^\infty g_n(u) = \mu,$$

a result proved by Khintchine (1955) and McFadden (1962).

The next step is to identify the function corresponding to the ordinary renewal density. We first confine our attention to regular processes. The second-order product density in the stationary regime can be described by the limiting behaviour of $h_2(t_1, t_2)$ $(t_2 > t_1)$ as the origin is shifted to $-\infty$. Thus we have

$$(4.3.12) \quad h_2(x_1, x_2) = \mu h_1^c(x_2 - x_1) \quad (x_2 > x_1),$$

where $h_1^c(.)$ is the conditioned product density of degree one defined by

$$(4.3.13a) \quad h_1^c(x) = \lim_{\substack{\Delta \to 0 \\ \Delta' \to 0}} \frac{\Pr\{N(x, \Delta) \geqslant 1 \mid N(0, \Delta') \geqslant 1\}}{\Delta}$$

or

$$(4.3.13b) \quad h_1^c(x) = \lim_{\substack{\Delta \to 0 \\ \Delta' \to 0}} \frac{E\{N(x, \Delta) \mid N(0, \Delta') = 1\}}{\Delta}$$

The conditioned product density is an intensity function and can be identified with $\lambda(x \mid \mathcal{H}_t)$ introduced in (4.3.2), using the information that $N(0, \Delta') = 1$ as $\mathcal{H}_t$. Thus the intensity function is uniquely specified by the history of the process confined to the interval $(t - \Delta', t)$. This limited history is sufficient for our purpose, in virtue of the homogeneity generated by the stationarity property of the univariate point process under consideration. We can regard the conditioned product density of degree one as the appropriate *generalization of the ordinary renewal density* since

$$(4.3.14) \quad \lim_{\Delta \to 0} E\{N(0, x) \mid N(0, \Delta) = 1\} = \int_0^x h_1^c(u) \, du.$$

Higher-order product densities are re-expressed in terms of conditioned product densities in a similar manner. The product density of degree m is given by

$$(4.3.15) \quad h_m(x_1, x_2, \ldots, x_m) = \mu h_{m-1}^c(x_2 - x_1, \ x_3 - x_1, \ldots, x_m - x_1)$$

$$(x_2 > x_1, x_3 > x_1, \ldots, x_m > x_1),$$

where $h_m^c(., ., ., \ldots, .)$ is the conditioned product density of degree m defined by

$$(4.3.16a) \quad h_m^c(u_1, u_2, \ldots, u_m)$$

$$= \lim_{\Delta \to 0} \lim_{\Delta_1 \to 0} \lim_{\Delta_2 \to 0} \ldots \lim_{\Delta_m \to 0} \Pr\{N(u_1, \Delta_1) \geq 1, N(u_2, \Delta_2) \geq 1,$$

$$\ldots N(u_m, \Delta_m) \geq 1 \mid N(0, \Delta) \geq 1\}/\Delta_1 \Delta_2 \ldots \Delta_m, \quad u_1 \neq u_2 \neq \ldots \neq u_m,$$

or

$$(4.3.16b) \quad h_m^c(u_1, u_2, \ldots, u_m)$$

$$= \lim_{\Delta \to 0} \lim_{\Delta_1 \to 0} \lim_{\Delta_2 \to 0} \ldots \lim_{\Delta_m \to 0} E[N(u_1, \Delta_1) N(u_2, \Delta_2)$$

$$\ldots N(u_m, \Delta_m) \mid N(0, \Delta) \geq 1]/\Delta_1 \Delta_2 \ldots \Delta_m, \quad (u_1 \neq u_2 \neq \ldots \neq u_m).$$

For instance, the mean square value of $N(t, x)$ which is the same as that of $N(0, x)$ (the origin being an arbitrary point) is given by

$$(4.3.17) \quad E[N(0, x)]^2 = \int_0^x \int_0^x h_2(x_1, x_2) \, dx_1 dx_2 + \int_0^x h_1(x_1) \, dx_1.$$

Using the implications of stationarity, we find that

$$\mathrm{E}\,[N(0,x)]^2 = \mu x + 2\mu \int\limits_0^x dx_1 \int\limits_{x_1}^x h_1^c(u-x_1)\,du$$

(4.3.18)
$$= \mu x + 2\mu \int\limits_0^x (x-u)\,h_1^c(u)\,du.$$

In a similar way higher-order moments can be calculated.

Thus in the limited context of regular point processes, we can regard the product densities $h_1^c(.), h_2^c(.,.), \ldots$ as the *generalized renewal densities*; in general, $h_2^c(.,.)$ and higher-order densities cannot be expressed as a product of $h_1^c(.)$, as is to be expected.

The interpretation of conditioned product densities as generalized renewal densities no longer holds good for non-regular processes. There are several reasons as to why the passage to general stationary processes is not smooth. The set of equations (4.3.16a) is not equivalent to the set (4.3.16b). Thus the densities themselves must be very carefully defined. For instance, definition (4.3.16a) for non-regular processes merely implies that the right-hand side is the probability density of at least one event in each of the locations $x_1, x_2, \ldots, x_m$, irrespective of the presence of events in other locations, and this is by no means sufficient to determine the moments of the number of events, since it can provide only a lower bound to the moments. On the other hand, equation (4.3.16b) deals with the correlation of the density of events, and we shall call the densities corresponding to (4.3.16b) the *product densities* since they still represent the *expected value of the product of densities of events*. There is also another reason why (4.3.16b) is the appropriate definition of the product density of irregular point processes. As we have seen in Chapter 2, the product densities of regular point processes arise as functional derivatives of the characteristic functional evaluated at appropriate points. The definition implied by (2.4.11) is equally applicable to irregular point processes, provided the arguments $x_1, x_2, \ldots, x_m$ are distinct.

To obtain the moment formulae similar to (2.3.15) or (2.3.16), we note that the product densities as defined by (4.3.16b) are not convenient quantities to start with. This is due to the peculiar manner in which they arise for distinct values of the different arguments. In the case of regular processes, the value at $u_i = u_j$ is defined to be the limit as $u_i \to u_j$, the limit being unique by virtue of regularity. However, in the present case we have to modify definition (4.3.16b) by relaxing the condition $u_1 \neq u_2 \neq \ldots \neq u_n$, so that a multiple event occurring at u_i will be represented by the density function in which the argument repeats as many times as the degree of its multiplicity. Since the h-function is essentially a density, it necessarily contains delta functions at the points at which multiple events occur. Thus we have two different density functions defined by (4.3.16b), according as to whether the condition $u_1 \neq u_2 \neq \ldots \neq u_n$ is imposed or is not imposed on the right-hand side. In the latter case we shall call the sequence of densities the *factorial moment densities* since product moments can be obtained by integration over the appropriate range.

To make further progress, it is convenient to deal with the product densities. For instance, the mean square number of events that occur in an interval $(t, t + x)$ (t being an arbitrary point) is given by

$$(4.3.19) \quad E\,[N(t, x)]^2 = \int\limits_{u=t}^{t+x} \int\limits_{v=t}^{t+x} E\,[d_uN(0, u)d_vN(0, v)]$$

$$= \int\limits_{u=t}^{t+x} E\,[d_uN(0, u)]^2 + \int\limits_{u=t}^{t+x} \int\limits_{v=t}^{t+x} E\,[d_uN(0, u)d_v\,[N(0, v)]\,, \quad u \neq v,$$

or equivalently, using stationarity,

$$(4.3.20) \quad E\,[N(t, x)]^2 = x\,\Sigma\,i^2\lambda_i + 2\mu \int\limits_0^x (x - u)\,h_1^c(u)\,du,$$

where $h_1^c(.)$ is defined for positive definite values of its argument by

$$(4.3.21) \quad h_1^c(u) = \lim_{\Delta \to 0} E\,[d_uN(0, u)|N(-\Delta, \Delta) \geqslant 1]/du \quad u > 0,$$

$$(4.3.22) \quad \lambda_i = F_i(0),$$

$$(4.3.23) \quad F_i(x) = \lim_{\Delta \to 0} \{\text{Probability that the } i\text{th event counted from the origin occurs earlier than } x\,|\,N(-\Delta, \Delta) \geqslant 1\}$$

$$= \lim_{\Delta \to 0} \Pr\,\{N(0, x) \geqslant i\,|\,N(-\Delta, \Delta) \geqslant 1\}.$$

Comparing (4.3.20) with the corresponding result from renewal theory (of regular stationary processes) given by equation (3.4.9), we conclude that $h_1^c(.)$ cannot be considered as the appropriate generalization of the renewal density. However, if we add to the continuous part of the density $h_1^c(.)$ the appropriate delta function contribution so as to render (4.3.19) valid when u tends to zero from the right, we arrive at the *generalized renewal density*, namely

$$(4.3.24) \quad r_1(u) = h_1^c(u) + \frac{1}{2\mu}\,\Sigma\,k(k - 1)\lambda_k\,\delta(u).$$

In order to obtain a generalization of renewal densities of higher order, we note that the form given by (4.3.24) can be arrived at by analogy with equation (4.3.15). Thus the generalized renewal density is simply the stationary second factorial moment density (with its arguments ordered) divided by twice the stationary value of the average intensity. This argument can be extended to any order. For instance, the generalized renewal density of order $n - 1$ is simply the stationary factorial moment density with ordered arguments of order n divided by $n!$ times the stationary value of the average intensity, the factor $n!$ arising from the different possible orderings of the arguments of the factorial moment density (which is by definition a symmetric function of its arguments):

(4.3.25) $r_{n-1}(x_1, x_2, \dots, x_{n-1})$

$$= \frac{1}{\mu n!} \lim_{\Delta \to 0} E\left[\prod_{i=1}^{n-1} d_{x_i} N(0, x_i) \mid N(-\Delta, \Delta) \geqslant 1\right] \prod_{i=1}^{n-1} dx_i \quad (0 < x_1 < x_2 < \dots x_{n-1})$$

Thus we can conclude that the factorial moment densities are the convenient quantities to handle. The Fourier transform of the second-order density yields the power spectrum which forms the basis for the spectral analysis of the process. It should, however, be noted that in any particular process it may not be possible to arrive at the factorial moment densities, particularly if there are correlations between different multiplicities at different locations, as, for instance, in the growth of a population process with twins and higher multiplets. In such a case the multiple-point product densities introduced in Chapter 2 must be used.

4.4   Moment formulae

We now proceed to show how the moments and correlations of the random variable $N(0, x)$ can be obtained with the help of the densities introduced in section **4.3**. The second-order correlation of $N(0, x)$ is given by

(4.4.1) $\displaystyle E[N(0, x)N(0, y)] = \int_0^x \int_0^y E[d_u N(0, u) d_v N(0, v)]$

$$= \int_0^x du \int_0^y h_2(u, v) \, dv + \int_0^{\min(x, y)} E[d_v N(0, v)]^2.$$

Using (4.3.19), we find that

(4.4.2) $E[N(0, x)N(0, y)]$

$$= \mu \int_0^x du \int_u^y h_1^c(v - u) \, dv + \mu \int_0^{\min(x, y)} du \int_0^u h_1^c(u - v) \, dv + v \min(x, y),$$

where

(4.4.3) $v = \Sigma i^2 \lambda_i.$

If we put $x = y$, we obtain the moment formula

(4.4.4) $\displaystyle E[N(0, x)]^2 = 2\mu \int_{0^+}^x (x - u) h_1^c(u) \, du + vx$

$$= \mu x + 2\mu \int_{0^-}^x (x - u) r_1(u) \, du.$$

The above result can also be obtained with the help of the sequence of functions $G_n(.)$ (see problem 4.12, below).

We can also evaluate the correlation of the number of events in any two consecutive non-overlapping intervals. This is given by

(4.4.5) $E[N(0, x)N(x, y)] = E[N(0, x)N(0, y)] - E[N(0, x)]^2.$

Using equations (4.4.2) and (4.4.4), we find after some calculation

(4.4.6) $E[N(0, x)N(x, y)] = \mu \left[\int_0^{x+y} H_1(u)\, du - \int_0^x H_1(u)\, du - \int_0^y H_1(u)\, du \right],$

where $H_1(.)$ is the generalized renewal function. Equation (4.4.6) was derived by McFadden (1962) by an indirect method, using the properties of $G_n(.)$.

The second-order properties of the stationary point process viewed as a counting process are also readily obtained. We start with the formula

(4.4.7) $E[d_x N(0, x)\, d_y N(0, y)] = [\mu h_1^c(y-x) + \delta(x-y)\nu]\, dx\, dy \quad x \leqslant y.$

The covariance density is given by

(4.4.8) $\gamma(u) = \mu h_1^c(|u|) + \delta(u)\nu - \mu^2,$

where $\gamma(.)$ is defined for both positive and negative values of u. The spectral density is given by

(4.4.9) $\tilde{\gamma}(\omega) = \dfrac{\nu}{2\pi} + \dfrac{\mu}{2\pi} \int_{-\infty}^{+\infty} (h_1^c(|u|) - \mu)\, e^{-i\omega u}\, du.$

The properties of $\tilde{\gamma}(\omega)$ for regular stationary processes have been investigated in great detail and an excellent account of these can be found in the monograph of Cox and Lewis (1966).

Next we proceed to connect the generalized renewal densities to the probability generating functional of $N(0, x)$. To do this we note that the mth factorial moment is given by

(4.4.10) $E[N(0, x)]_m = \int_0^x \int_0^x \ldots \int_0^x E\left[\prod_{i=1}^m d_{x_i} N(0, x_i) \right].$

Using stationarity and the symmetric nature of the integrand, we obtain

(4.4.11) $E[N(0, x)]_m$

$= m \int_0^x dx_1 \int_{x_1}^x \ldots \int_{x_1}^x \lim_{\Delta \to 0} E\left[\prod_{i=2}^m d_{x_i} N(0, x_i - x_1) \mid N(-\Delta, \Delta) \geqslant 1 \right] / \Delta.$

Using (4.3.23), we find that

(4.4.12) $E[N(0, x)]_m$

$= m! \mu \int_0^x dx_1 \int_0^{x-x_1} dx_2 \int_{x_2}^{x-x_1} \ldots \int_{x_{m-1}}^{x-x_1} dx_m\, r_{m-1}(x_2, x_3, \ldots, x_m).$

Setting

(4.4.13) $r_m(x_1, x_2, \ldots, x_m) = T_m(x_1, x_2 - x_1, x_3 - x_2, \ldots, x_m - x_{m-1})$

and using the definition of the generating function of $N(0, x)$, we obtain

(4.4.14) $\phi(z, x) = \mathrm{E}\,[z^{N(0,\,x)}]$

$$= 1 + \sum_{1}^{\infty} (z - 1)^m \, \mu \int_0^x dx_1 \int_{x_1}^x dx_2 \ldots \int_{x_{m-1}}^x dx_m T_{m-1}(x_2, x_3 - x_2, \ldots, x_m - x_{m-1}),$$

where $T_0 \equiv 1$. The Laplace transform with respect to x is given by

(4.4.15) $\phi^*(z, s) = \int_0^\infty \phi(z, x)\, e^{-sx}\, dx$

$$= \frac{1}{s} + \frac{\mu(z - 1)}{s^2} + \frac{\mu(z - 1)}{s^2} \sum_{m=1}^\infty (z - 1)^m \, T_m^*(s, s, \ldots, s),$$

a result first obtained by McFadden and Weissblum (1963). In a similar way the Laplace transform of the bivariate generating function of $N(0, .)$ can be connected to $r^*(\ldots)$. However, in practical computation it is very difficult to obtain $r(\ldots)$ for general processes, and it is in this context that the formalism of multiple product densities is useful.

For purposes of illustration, let us consider a generalized Poisson process (see Parzen (1962), Chapter 4) which is also known as a "quasi-Poisson" process (see Smith (1958)). The process arises by assigning multiplicities which are independently and identically distributed. Following McFadden (1962), we call the basic events locations, multiple events being assigned to each one of these locations. Let $b > 0$ be the parameter characterizing the Poisson process of locations, and let $g(.)$ be the probability generating function governing the multiplicities of events at any location. Then the following relations are obvious:

(4.4.16) $b = \lambda, \quad bg'(1) = \mu,$

so that $b/\mu < 1$. From section 2.6 it follows that the process is a Poisson cluster process, and its characteristic functional is seen to be

$$\Phi(\theta, T) = \mathrm{E}\left\{ \exp i \int_0^T \theta(t)\, d_t N(0, t) \right\}$$

(4.4.17) $$= \exp -\left[b \int_0^T [1 - g(e^{i\theta(t)})]\, dt \right].$$

The product densities (in the formulation of multiple point densities of section 2.5) are given by

(4.4.18) $h_1(t) = \Sigma\, i\, \lambda_i$

$$= bg'(1) = \mu,$$

$$(4.4.19) \quad h_2(t_1, t_2) = [bg'(1)]^2 = \mu^2 \quad (t_1 \neq t_2).$$

The corresponding generalized renewal densities can be directly evaluated using combinatorial arguments:

$$(4.4.20) \quad r_1(x) = \mu + \frac{b}{2\mu} g''(1) \delta(x)$$

$$(4.4.21) \quad r_2(x, y) = \mu^2 + \frac{b}{6\mu} g'''(1) \delta(x) \delta(y)$$

$$+ \frac{b^2}{2\mu} g'(1) g''(1) [\delta(x) + \delta(x - y)].$$

There are some interesting special cases of the generalized Poisson process. Suppose we assign multiplicities by Bernoulli trials, so that $g(.)$ is given by

$$(4.4.22) \quad g(z) = \frac{b}{\mu} z \bigg/ \left[1 - \left(1 - \frac{b}{\mu}\right) z \right].$$

The process is known as the "modified Poisson process". We next observe that the characteristic functional of the process is given by

$$(4.4.23) \quad \Phi(\theta, T) = \exp - \left\{ b \int_0^T \left[1 - \frac{b}{\mu} \frac{e^{i\theta(t)}}{1 - \left(1 - \frac{b}{\mu}\right) e^{i\theta(t)}} \right] dt \right\}.$$

Putting $e^{i\theta(t)} = $ a constant $= z$, we obtain the generating function of the counting process:

$$(4.4.24) \quad \phi(z, T) = \exp \left[-\frac{b \mu T(1 - z)}{\mu - (\mu - b)z} \right].$$

This result was obtained by McFadden (1962). Since the Bernoulli trials are independent, the point process is an "irregular renewal" process. (The process can be made regular by putting $b = \mu$, in which case it reduces to a simple Poisson process.) The interval between any two successive events is governed by the probability density function $f(.)$ given by

$$(4.4.25) \qquad\qquad f(u) = \left(1 - \frac{b}{\mu}\right) \delta(x) + \frac{b^2}{\mu} e^{-bx},$$

so that

$$(4.4.26) \qquad\qquad F(0) = 1 - \frac{b}{\mu} < 1.$$

It should be noted that the choice of $g(.)$ corresponding to the Bernoulli distribution converts the process into a renewal process. Otherwise the intervals between events are not independently distributed, although the locations are.

70

It is interesting to investigate whether the irregular renewal process defined by (4.4.23) is the most general one. The answer is in the negative, since we can consider locations which form a renewal process, multiplicities being assigned to these locations as before. It is instructive to derive the generalized renewal densities in this case. Let the multiplicity "n" be assigned to each location, with probability λ'_n, so that $\sum\limits_{n=1}^{\infty} \lambda'_n = 1$. Then the first-order product density is given by

$$(4.4.27) \qquad h_1(t) = \mu' \sum i\lambda'_i = \mu,$$

where μ' is the stationary value of the product density of degree one of locations. The generalized renewal densities of order one and two are given by

$$(4.4.28) \quad r_1(x) = h(x) \sum i\lambda'_i + \frac{1}{2\mu} \sum i(i-1)\lambda'_i \delta(x),$$

$$(4.4.29) \quad r_2(x_1, x_2)$$

$$= h(x_1)(\sum i\lambda'_i)^2 h(x_2) + \frac{1}{6\mu} \sum i(i-1)(i-2)\, \lambda'_i \delta(x_1)\delta(x_2)$$

$$+ \frac{1}{2\mu} \sum i(i-1)\lambda'_i \{h(x_1)\delta(x_2-x_1) + h(x_2-x_1)\delta(x_1)\}(\sum i\lambda'_i) \quad (x_1 < x_2),$$

where $h(.)$ is the ordinary renewal density of the locations.

The process becomes a renewal process if

$$(4.4.30) \quad \lambda'_n = \frac{\mu'}{\mu}\left(1 - \frac{\mu'}{\mu}\right)^{n-1} \quad (n \geq 1, \mu' \leq \mu),$$

in which case the interval distribution is specified by

$$(4.4.31) \quad f_p(x) = \left(1 - \frac{\mu'}{\mu}\right)\delta(x) + \frac{\mu'}{\mu} f(x),$$

where the subscript p distinguishes the density function of the interval governing the points of the point process from that corresponding to the locations.

4.5 Some special point processes

We consider in this section some point processes that have been studied in the literature with special reference to certain concrete applications. Since these processes illustrate many of the general points that have been discussed earlier, we have found it worthwhile to discuss some of the processes in detail.

(i) Wold's Markov process of intervals

As a generalization of renewal processes, Wold (1948a, b) considered point processes having after-effect or local contagion. He assumed that the sequence of intervals $y_1, y_2, \ldots$ forms a homogeneous Markov chain, so that the stationary distribution of the sequence of intervals can be described by the density function $f(x|y)$, where

(4.5.1) $\quad f(x|y) = \lim_{\Delta \to 0} \Pr\{x \leqslant y_j \leqslant x + \Delta \,|\, y_{j-1} = y\}.$

The marginal probability density function $\mathcal{F}_j(.)$ of y_j satisfies the relation

(4.5.2) $\qquad\qquad \mathcal{F}_j(x) = \int_0^\infty f(x|y)\, \mathcal{F}_{j-1}(y)\, dy.$

If a stationary process exists, then the marginal distribution of intervals becomes independent of j, and denoting the corresponding density function by $f_1(.)$, we find that $f_1(.)$ is the solution of the equation

(4.5.3) $\qquad\qquad f_1(x) = \int_0^\infty f_1(y)\, f(x|y)\, dy.$

The above equation is, for general $f(.\,|.)$, intractable, and Wold (1948a) restricted himself to an exponential distribution of the form

(4.5.4) $\qquad\qquad f(x|y) = \lambda(y)\, e^{-\lambda(y)x}.$

If we choose $\lambda(y)$ as

(4.5.5) $\qquad\qquad \lambda(y) = \lambda y^{-1/2},$

the probability density function $f_1(.)$ can be solved by the Mellin transform technique. If $\psi(s)$ is the Mellin transform, then we obtain from (4.5.3) and (4.5.4)

$$\psi(s) = \int_0^\infty x^{s-1} f_1(x)\, dx$$

(4.5.6) $\qquad\qquad\qquad = \Gamma(s+1)\, \psi(s/2).$

Solving the equation recursively, we find that

(4.5.7) $\qquad\qquad \psi(s) = \prod_0^\infty \Gamma\left(1 + \frac{s}{2^n}\right).$

Using the meromorphic nature of the gamma function, we note that the solution can be written in the form

(4.5.8) $\qquad\qquad f_1(x) = \sum_{n=0}^\infty x^n \sum_{\nu=0}^n A_{n\nu}\, (\log x)^\nu.$

The first few coefficients were estimated numerically by Wold:

(4.5.9) $\quad \begin{cases} A_{00} = 2\cdot555, \quad A_{10} = 2\cdot9803, \quad A_{11} = 5\cdot11 \\ A_{20} = -11\cdot9362, \quad A_{21} = A_{22} = 0. \end{cases}$

The series is found to converge very rapidly.

Another special case corresponding to

(4.5.10) $$\lambda(y) = \lambda_0(1 + \lambda_1 y),$$

where λ_1 is a small parameter, has been dealt with by Cox (1955). In such a case $f_1(x)$ is given by

(4.5.11) $$f_1(x) = \lambda_0\, e^{-\lambda_0 x}\left\{\lambda_0 + \lambda_1\left(\frac{1}{\lambda_0} - x\right)\right\} + o(\lambda_1).$$

For positive λ_1, it is clear that a large interval is followed by an interval with a small mean and that the successive intervals are negatively correlated. More recently, Lampard (1968) has obtained a solution of equation (4.5.3) when $f(x|y)$ is closely related to the modified Bessel function of the first kind (see problem 4.15, below).

We next attempt to provide a description of the point process in terms of the product densities. Let us assume that the process is regular. Introducing the product density of degree one conditioned at the origin by

(4.5.12) $$h_1(t, x) = \lim_{\Delta_1, \Delta_2, \Delta_3 \to 0} \Pr\{N(t, \Delta_1) = 1 | N(0, \Delta_2) = 1,$$

$$N(-x + \Delta_3, x - \Delta_3) = 0,\ N(-x, \Delta_3) = 1\}/\Delta_1,$$

we find, using the stationary Markov nature of the successive intervals, that

(4.5.13) $$h_1(t, x) = f(t|x) + \int_0^\infty f(y|x)\, h_1(t - y, y)\, dy.$$

Defining the double Laplace transform of $h_1(t, x)$ as $h_1^*(s, p)$, we find that

(4.5.14) $$h_1^*(s, p) = f^*(s|p) + \frac{1}{2\pi i}\int_{\sigma - i\infty}^{\sigma + i\infty} f^*(\alpha|p)\, h_1^*(s, s - \alpha)\, d\alpha \qquad 0 < \sigma < \mathrm{Re}\ s$$

where $f^*(s|p)$ is the double Laplace transform of $f(t|x)$. If $f^*(.|.)$ is given, equation (4.5.14) can be solved (see problem 4.16, below).

(ii) *Delayed renewal processes*

A non-renewal point process can be generated by subjecting each event of a renewal point process to a random delay. For simplicity, let us assume that the delays are independently and identically distributed with probability density function $\alpha(.)$. Blanc-Lapierre *et al.* (1965) studied a particular model in which Poisson events are subject to such delays. Let us assume that the renewal process is specified by the renewal density function $h(.)$. It is easy to derive explicit expressions for the product densities of the first few orders of delayed events. Using the superscript d to distinguish the product densities of delayed events, it follows by elementary arguments that

(4.5.15) $h_1^d(x_1) = \int_0^\infty dy_1 \, h_1(x_1 - y_1)\alpha(y_1) = \mu$

(4.5.16) $h_2^d(x_1, x_2) = \int_0^\infty \int_0^\infty h_2(x_1 - y_1, \, x_2 - y_2)\alpha(y_1)\alpha(y_2)\,dy_1 dy_2$

$$= \mu \int_0^\infty dy_1 \int_0^{y_1 - x_1 + x_2} h(x_2 - x_1 - y_2 + y_1)\alpha(y_1)\alpha(y_2)\,dy_2$$

$$+ \int_0^\infty dy_1 \int_{y_1 - x_1 + x_2}^\infty h(x_1 - x_2 - y_1 + y_2)\alpha(y_1)\alpha(y_2)\,dy_2,$$

where μ is the stationary value of the first-order product density of the renewal process and $h_1(.)$ and $h_2(.,.)$ are the product densities of the primary events. In principle it is possible to write down explicit expressions for higher-order product densities. Delayed renewal processes have also been studied by ten Hoopen and Reuver (1967c) in connection with models for neuronal firing.

(iii) A simple non-renewal process

A simple non-renewal process can be generated by defining the corresponding interval lengths y_i by

(4.5.17) $$y_i = \sigma_{i-1} + \sigma_i,$$

where the $\{\sigma_i\}$ are the interval lengths of a stationary renewal point process. It is to be noted that even if the σ-intervals are exponentially distributed, the point process generated by the y-intervals is not a renewal process. The point process is also not of the Wold type discussed in (i) above. We shall indicate how the product density of degree n can be explicitly obtained, thus providing a complete characterization of the point process.

Let us call the σ_i "the associated σ-intervals corresponding to the intervals y_i". We shall denote the σ-interval corresponding to the point x by σ_x. It is convenient to introduce the conditioned product density

(4.5.18) $h_1^c(x, u) = \lim_{\Delta_1, \Delta_2, \Delta_3 \to 0} \Pr\{$a y-event occurs in $(x, x + \Delta_1)$,

$u < \sigma_x < u + \Delta_2 |$ a y-event in $(-\Delta_3, 0)\}/\Delta_1\Delta_2$.

We note that although the y-events are not regenerative, additional information specifying the length of σ_x renders the point x regenerative. It is easy to see that $h_1^c(x, u)$ satisfies the equation

(4.5.19) $$h_1^c(x, u) = \frac{h_1(\infty, x - u)\phi(u)}{h_1(\infty)} + \int_{\frac{x-u}{2}}^{x-u} h_1^c(v, x - u - v)\phi(u)\,dv,$$

where $h_1(\infty)$ is the stationary product density of degree one of y-events.

74

We next observe that

$$(4.5.20) \qquad h_1(\infty, u) = h_1(\infty)\phi(u).$$

Incidentally $h_1(\infty)$ can also be evaluated:

$$(4.5.21) \quad 1/h_1(\infty) = \int_0^\infty u \, du \int_0^u \phi(v)\phi(u-v) \, dv.$$

Using (4.5.20) in (4.5.19), we obtain

$$(4.5.22) \quad h_1^c(x, u) = \phi(u) \left[\phi(x-u) + \int_{(x-u)/2}^{x-u} h_1^c(v, x-u-v) \, dv \right].$$

Observing that $h_1^c(x, u)/\phi(u)$ is a function of $x-u$ only, we set

$$(4.5.23) \qquad h_1^c(x, u) = \phi(u) \hbar_1(x-u)$$

and obtain

$$\hbar_1(x) = \phi(x) + \int_{x/2}^x \hbar_1(2v-x)\phi(x-v) \, dv$$

or

$$(4.5.24) \quad \hbar_1(x) = \phi(x) + \int_0^{x/2} \hbar_1(x-2v)\phi(v) \, dv.$$

Thus the Laplace transform $\hbar_1^*(s)$ of $\hbar_1(.)$ is given by

$$(4.5.25) \qquad \hbar_1^*(s) = \frac{\phi^*(s)}{1 - \phi^*(2s)}.$$

In the special case when the σ-intervals are exponentially distributed with parameter λ, we find that

$$(4.5.26) \qquad \hbar_1(u) = \frac{\lambda}{2}(1 + e^{-\lambda u})$$

$$(4.5.27) \qquad h_1^c(x, u) = \frac{\lambda^2}{2}(e^{-\lambda u} + e^{-\lambda x}),$$

so that the conditioned first-order product density of y-events is given by

$$(4.5.28) \qquad h_1^c(x) = \int_0^x h_1^c(x, u) \, du$$

$$= \frac{\lambda}{2}[1 - e^{-\lambda x} + \lambda x \, e^{-\lambda x}].$$

The limit of $h_1^c(x)$ as x tends to infinity is $\lambda/2$, as is to be expected. If $N_y(t, x)$ is the number of y-events in $(t, t+x)$, we find, using formula (4.4),

$$(4.5.29) \qquad E\,[N_y(t, x)] = \frac{\lambda}{2}x$$

$$(4.5.30) \quad E\left[\{N_y(t,x)\}^2\right] = \frac{\lambda}{2}x + 2\int_0^x \frac{\lambda}{2}(x-u)h_1^c(u)\,du$$

$$= \frac{\lambda}{2}x + \frac{\lambda^2 x^2}{4} + \tfrac{1}{2}(e^{-\lambda x} + \lambda x - 1).$$

Returning to the general case, we note that the second-order product density $h_2^c(x_1, u_1; x_2, u_2)$ conditioned at the origin can also be obtained. Setting

$$(4.5.31) \quad h_2^c(x_1, u_1; x_2, u_2) = h_1^c(x_1, u_1)\mathcal{F}(x_2, u_2 \,|\, x_1, u_1),$$

we note that an integral equation for $\mathcal{F}$ can be set up, since the length of the σ-interval associated with any y-event renders the event regenerative. Thus we have

$$(4.5.32) \quad \mathcal{F}(x_2, u_2 \,|\, x_1, u_1)$$

$$= \phi(u_2)\,\delta(x_2 - x_1 - u_1 - u_2) + \phi(u_2)\int_{(x_1+x_2+u_1-u_2)/2}^{x_2-u_2} \mathcal{F}(v, x_2 - v - u_2 \,|\, x_1, u_1)\,dv$$

or equivalently

$$(4.5.33) \quad \mathcal{F}(x_2, u_2 \,|\, x_1, u_1)$$

$$= \phi(u_2)\,\delta(x_2 - x_1 - u_1 - u_2) + \phi(u_2)\int_0^{(x_2-x_1-u_1-u_2)/2} \mathcal{F}(x_2 - v - u_2, v \,|\, x_1, u_1)\,dv.$$

The above equation shows that $\mathcal{F}(x_2, u_2 \,|\, x_1, u_1)/\phi(u_2)$ is a function of the single argument $x_2 - x_1 - u_1 - u_2$. Making the substitution

$$(4.5.34) \quad \mathcal{F}(x_2, u_2 \,|\, x_1, u_1)/\phi(u_2) = \hbar_2(x_2 - x_1 - u_1 - u_2),$$

we obtain

$$(4.5.35) \quad \hbar_2(u) = \delta(u) + \int_0^{u/2} \hbar_2(u - 2v)\,\phi(v)\,dv.$$

Thus the Laplace transform $\hbar_2^*(s)$ of $\hbar_2(.)$ can be found to be

$$(4.5.36) \qquad \hbar_2^*(s) = \frac{1}{1 - \phi^*(2s)}.$$

Thus for any given $\phi(.)$, $h_2(.)$ can be determined. If $h_n^c(x_1, u_1; x_2, u_2; \ldots x_n, u_n)$ is the conditioned product density of degree n, then it follows easily that

$$(4.5.37) \quad h_n^c(x_1, u_1; x_2, u_2; \ldots x_n, u_n)$$

$$= h_1^c(x_1, u_1)\prod_{i=2}^n \mathcal{F}(x_i, u_i \,|\, x_{i-1}, u_{i-1}),$$

so that we have a complete characterization of the point process in terms of the product densities.

CHAPTER 4: PROBLEMS

4.1 Show that, for a renewal process, simple stationarity defined through the invariance of the distribution function of $L_n(t)$ implies complete stationarity.

4.2 A stationary process is said to be completely stationary in the forward (backward) sense if for each positive (negative) integer set $k_1, k_2, \ldots, k_n$ the multivariate distribution of $Lk_1(t), Lk_2(t), \ldots, Lk_n(t)$ is invariant under translation. Show that complete stationarity in the forward or backward sense is equivalent to complete stationarity of the counting process as defined by (4.1.3).

4.3 Find the joint distribution function of L_1 and L_{-1} for a stationary point process. Show that L_1 and L_{-1} are independent if $p(0,.)$ (see (4.2.7)) is exponential.   (McFadden and Weissblum, 1963)

4.4 Prove directly that L_1 and L_{-1} are identically distributed provided the process is stationary.   (Cox, 1962)

4.5 Show that $\phi^*(z, s)$ satisfies the relation

$$\phi^*(z, s) = \frac{1}{s} - \frac{1-z}{\mu_1 s^2} + \frac{(1-z)^2}{\mu_1 s_z^2}\, \theta_f^*(z, s)$$

$$= \frac{1}{s} + \frac{z-1}{sz}\, \theta_g^*(z, s),$$

where $\theta_f(z, x)$ and $\theta_g(z, x)$ are the generating functions of $g_n(.,x)$ and $f_n(.,x)$:

$$\theta_f(z, x) = \sum_1^\infty z^n f_n(.,x)$$

$$\theta_g(z, x) = \sum_1^\infty z^n g_n(.,x).$$ (McFadden, 1962)

4.6 Orderliness can be defined in terms of the distribution $G_n(.)$ (or $F_n(.)$): a point process is orderly if $G_n(\Delta)/G_1(\Delta) = O(\Delta)$, $n > 1$. Show that this definition is equivalent to that given in (4.3.1).

4.7 Show that stationarity implies that

$$\mathrm{E}\,[N(t, x)]^k = O(x^k) \quad \text{for } x \to \infty.$$

Consider the Yule–Furry process (section 2.5) and show that $\mathrm{E}\,[N(t, x)]^k$ increases faster than any power of x as x tends to infinity.

4.8   The random variable χ_n is defined as

$$\chi_n(x) = 1 \quad \text{if } N(t, x) = n$$
$$= 0 \quad \text{otherwise.}$$

Show that the expected value of the time spent in state n is given by

$$\text{E}\left[\int_0^\infty \chi_n(x)\, dx\right] = \int_0^\infty p(n, t)\, dt.$$

4.9   Show that

$$\int_0^\infty t\, p(n, t)\, dt = \frac{\mu}{2} \text{E}\{y_0 y_n [y_0 + 2(y_1 + y_2 \dots + y_{n-1}) + y_n]\}.$$

(McFadden, 1962)

4.10   Show that

$$\int_0^\infty t^n p(1, t)\, dt = \frac{\mu}{(n+1)(n+2)} \text{E}\{(y_0 + y_1)^{n+2} - y_0^{n+2} - y_1^{n+2}\}.$$

4.11   Prove that $\sum_{n=1}^{\infty} G_n(x) = \mu x$ is a consequence of stationarity.

4.12   Use (4.4.5) to evaluate $\text{E}\,[N(0, x)]^2$ and deduce (4.4.4).

(McFadden, 1962)

4.13   *Quasi-Markov chain of intervals of zero and finite length*

Assume that in a point process, $\text{Pr}\,\{y_i = 0 \text{ or} > 0\}$ depends only on whether $y_{i-1} = 0$ or > 0. Let the non-zero interval lengths be distributed exponentially. This is equivalent to choosing a generalized Poisson process with

$$\lambda_1' = 1 - v, \quad \lambda_n' = uv(1-u)^{n \perp 2} \quad (n \geq 2, 0 < u < 1, 0 < v < 1).$$

Find the generating function of the point process and deduce the modified Poisson process as a special case.   (McFadden, 1962)

4.14   Verify that when (4.4.30) is imposed on (4.4.29),

$$r(x_1, x_2) = r(x_1)\, r(x_2 - x_1).$$

4.15   Show that if $f(x|y)$ is given by

$$f(x|y) = \beta\, e^{-\alpha x - \beta y} \left[\frac{\beta y}{\alpha x}\right]^{r/2} I_r\, [2(\alpha\beta xy)^{1/2}] \quad (\alpha < \beta),$$

then the unique solution of equation (4.5.3) is given by

$$f_1(x) = (\beta - \alpha) \frac{[(\beta - \alpha)x]^r}{r!}\, e^{-(\beta - \alpha)x}.$$

(Lampard, 1968)

78

4.16   Choosing the special form (4.5.11), show that for small λ_1, the solution of (4.5.14) can be written as

$$h_1^*(s \mid p) = \frac{\lambda_0}{(s+\lambda_0)p} + \frac{s\lambda_0\lambda_1}{(s+\lambda_0)^2 p^2} + \alpha^*(s)\left(\frac{\lambda_0}{p} + \frac{\lambda_0\lambda_1}{p^2}\right) - \beta^*(s)\frac{\lambda_0^2\lambda_1}{p^2} + o(\lambda_1),$$

where

$$\alpha^*(s) = h_1^*(s, s+\lambda_0)$$

$$\beta^*(s) = \frac{\partial}{\partial p} h_1^*(s, p)|_{p=s+\lambda_0}.$$

By evaluating both sides of the above equation at $p = s + \lambda_0$ and differentiating both sides at the point $p = s + \lambda_0$, determine $\alpha^*(.)$ and $\beta^*(.)$ and hence $h_1(t)$ explicitly.

4.17   From (4.5.25) and (4.5.36) obtain $h_1(.)$ and $h_2(.)$ explicitly when

$$\phi(x) = (\lambda x)^{n-1}\frac{e^{-\lambda x}}{(n-1)!}\lambda$$

for $n = 2, 3$ and 4. Hence derive explicit expressions for the product density of degree m of y-events.

5 DOUBLY STOCHASTIC POISSON PROCESSES

5.1 Introduction

Let us consider an inhomogeneous (time dependent) Poisson point process with parameter $\lambda(t)$. If we assume that $\lambda(t)$ is itself the sample function of a stationary continuous stochastic process $\Lambda(t)$, then the process is known as a "doubly stochastic process". The terminology is due to Cox (1955) who studied the stoppages in a loom due to weft breaks. There are a number of instances in physics where problems of similar type arise. In the theory of space-charge limited shot noise, the number of pulses that occur in a given interval of time obey a doubly stochastic process, since the "Poisson" parameter λ undergoes random changes due to the individual pulses. In the theory of optical communications, the output of wide-band photodetectors is generally a shot noise process, the intensity parameter being identified as the intensity of the radiation impinging on the photodetector. Since generally the intensity is a randomly fluctuating quantity, we encounter a doubly stochastic Poisson process. There are two ways of looking at the resulting point process. The first, and straightforward way, consists in identifying it as a stationary point process whose intensity and other characteristics are related to the correlational structure of the stochastic process $\Lambda(t)$. This viewpoint may be useful in general situations. However, there is a slightly different viewpoint which makes use of the fact that the Poisson nature of the events for a given sample path $\Lambda(t)$ must have its implications on the correlational structure of the counting process. For instance, if we deal with the probability generating function of $N(0, T)$, i.e. the number of events in the interval $(0, T)$ (the origin being an arbitrary point), then the function can be identified as the expected value of $u^{N(0, T)}$ with respect to $\Lambda(.)$. Thus $g(u, T)$, the p.g.f., is given by

$$(5.1.1) \qquad g(u, T) = E\left[u^{N(0, T)}\right]$$

$$= E_{\Lambda(.)}\left[\exp (u - 1) \int_0^T \lambda(t)\, dt\right]$$

or

$$(5.1.2) \qquad g(u, T) = E\left[\exp (u - 1) \int_0^T \Lambda(t)\, dt\right],$$

so that $g(u, T)$ can be obtained from the characteristic function of the integral of $\Lambda(.)$. It should be borne in mind that in view of $\Lambda(.)$ being interpreted as the intensity, it should be positive. Perhaps even stationarity can be relaxed, in which case we obtain an evolutionary point process. In this chapter

79

we shall consider some simple models of $\Lambda(.)$ for which explicit results relatir
to the distribution of the counting process can be obtained. This is particularly important since there is a general feeling (see, for example, Cox and
Lewis (1966), p. 182) that there are no non-trivial, continuous parametric
stochastic processes for which the characteristic function defined by (5.1.2)
is known explicitly.

5.2 Models with a purely discontinuous intensity

Let us assume that $\Lambda(.)$ is a purely discontinuous stationary stochastic
process of the Kolmogorov–Feller type (see, for example, Gnedenko (1958))
so that the probability density function $\pi(x, t|x_0)$ defined by

$$(5.2.1) \quad \pi(x, t|x_0) = \lim_{\Delta \to 0} \Pr\{x < \Lambda(t) < x + \Delta \mid \Lambda(0) = x_0\}/\Delta$$

satisfies the equation

$$(5.2.2) \quad \frac{\partial \pi(x, t|x_0)}{\partial t} = -\pi(x, t|x_0) \int_x \rho(x'|x)\,dx' + \int_x \pi(x', t|x_0)\rho(x|x')\,dx',$$

where

$$(5.2.3) \quad \rho(x'|x) = \lim_{\Delta \to 0} \pi(x', \Delta|x)/\Delta \quad (x' \neq x).$$

We shall further assume that the random variable $\Lambda(.)$ is non-negative with
probability one, so that the integration over x' on the right-hand side of
(5.2.2) is from 0 to ∞. Next we restrict $\rho(x'|x)$ to be a function of x' only,
in which case (5.2.2) becomes

$$(5.2.4) \qquad \frac{\partial \pi(x, t|x_0)}{\partial t} = -a\,\pi(x, t|x_0) + \phi(x),$$

where

$$(5.2.5) \qquad \rho(x|x') = \phi(x), \qquad a = \int_0^\infty \phi(x)\,dx.$$

Thus the quantity a can be interpreted as the total probability per unit t of
a transition from any state of the special discontinuous Kolmogorov–Feller
process. Such a model was used by Ramakrishnan (1954a) to represent the
density of interstellar matter. The model has been further studied by Srinivasa
and Vasudevan (1967a) who have shown that the random variable $\Lambda(t)$ –
$E[\Lambda(t)]$ approximates to an Uhlenbeck–Ornstein process (second order) for
a certain wide choice of $\rho(.)$, provided that $a \gg 1$.

The solution of (5.2.4) is given by

$$(5.2.6) \quad \pi(x, t|x_0) = \frac{\phi(x)}{a}(1 - e^{-at}) + \delta(x - x_0)\,e^{-at},$$

so that the stationary probability density function of $\Lambda(.)$ is given by $\phi(.)/a$.

We now make the approximation $a \gg 1$ and calculate the nth moment of $\int_0^T \Lambda(t)\,dt$:

$$(5.2.7) \quad E\left[\left(\int_0^T \Lambda(t)\,dt\right)^n\right] = \int_0^T \int_0^T \ldots \int_0^T E\left[\Lambda(t_1)\,\Lambda(t_2)\ldots\Lambda(t_n)\right]dt_1\,dt_2\ldots dt_n.$$

The nth order correlation of $\Lambda(.)$ can be evaluated using the stationary nature of $\Lambda(.)$, (5.2.6), and the approximation $a \gg 1$. Thus we find, after some calculation, that

$$(5.2.8) \quad E\left[\Lambda(t_1)\,\Lambda(t_2)\ldots\Lambda(t_n)\right]$$

$$= \int_{x_1}\int_{x_2}\ldots\int_{x_n} \psi(x_1)\,\psi(x_2)\ldots\psi(x_n)\,dx_1\,dx_2\ldots dx_n$$

$$\times \left\{1 + \frac{1}{a}\left[-\sum_{i=1}^{n}\delta(t_i - t_{i-1}) + \sum_{i=1}^{n}\frac{1}{\psi(x_i)}\delta(x_i - x_{i-1})\,\delta(t_i - t_{i-1})\right]\right\},$$

where

$$(5.2.9) \quad \psi(.) = \phi(.)/a$$

and the approximation $a \gg 1$ is made by substituting $\frac{1}{a}\delta(t)$ for e^{-at} (see Ramakrishnan (1954a)), $\delta(.)$ being the Dirac delta function. Thus we find

$$(5.2.10) \quad E\left[\int_0^T \Lambda(t)\,dt\right]^n = \bar{x}^n\,T^n + \frac{n(n-1)}{a}\bar{x}^{n-2}\,T^{n-1}\,(\text{var } x),$$

where the mean value $\bar{x}$ and var x are defined with reference to the stationary density function $\psi(.)$. Using (5.2.10), we find that the generating function of the doubly stochastic Poisson process is given by

$$(5.2.11) \quad g(u, T) = \left\{1 + \frac{(u-1)^2 T}{a}\,\text{var } x\right\}\exp\left[(u-1)\bar{x}\,T\right].$$

Thus the process is approximately Poisson with a correction factor of order $1/a$.

The interval between any two successive events has a density function $f(.)$, given by

$$(5.2.12) \quad f(t) = E_{\Lambda(.)}\left[\lambda(t)\exp - \int_0^T \lambda(u)\,du\right]$$

or

$$(5.2.13) \quad f(t) = -\frac{\partial}{\partial t}\,g(u, t)|_{u=0}$$

$$= e^{-\bar{x}t}\left(\bar{x} + \frac{\text{var } x}{a}(t\bar{x}+1)\right),$$

so that there is a positive inhibition of events. The stationary value of the product density of degree two of events can be directly calculated:

$$(5.2.14) \quad h_2(t_1, t_2) = E\left[\Lambda(t_1)\,\Lambda(t_2)\right]$$

$$= \bar{x}^2 + \operatorname{var} x \; e^{-a|t_1 - t_2|}$$

It should be noted that (5.2.14) is true even if the constraint $a \gg 1$ is dropped

Cox and Lewis (1966) dealt with the case when the model corresponding to $\Lambda(t)$ is a stationary Uhlenbeck–Ornstein process. However, there is an inherent difficulty in the model in that the sample functions $\lambda(t)$ can then become negative.

In reliability analysis we encounter a situation in which the average hazard rate of an equipment changes, depending on the type of performance demanded of it. For instance, when the equipment is operating actively, the hazard rate is certainly higher than when it is kept on standby. A model was proposed by Gaver (1963) who analysed the failure-time distribution of an equipment whose (average) hazard rate oscillates alternately between two specified values. Let the time-axis be divided by a sequence of time-points: $0 < t_1 < t_2 < \ldots t_n < t$. At times t_i ($i = 1, 2, \ldots, n$) environmental changes take place, so that the item in question is submitted in alternation to the two failure rates λ_1 and λ_2. Thus, during the (even) time-intervals $(0, t_1)$ $(t_2, t_3) \ldots$, the rate λ_1 prevails, while the rate λ_2 prevails in the (odd) time-intervals (t_1, t_2) (t_3, t_4), ... We assume that all the intervals (odd and even) are independently distributed, the odd and even intervals being characterized respectively by the probability density functions $\alpha(.)$ and $\beta(.)$.

It is useful to study the characteristics of the Poisson process so generated, the parameter of the Poisson process alternating between λ_1 and λ_2 in the manner prescribed above. The generating function governing the number of events in $(0, t)$ is given by

$$(5.2.15) \quad g(u, t) = E\left[\exp (u - 1) \int_0^t \lambda(v)\, dv\right],$$

where $\lambda(v)$ is the random variable alternating between the values λ_1 and λ_2. If we set $u = 0$ in (5.2.15) we obtain the probability of the survival of the equipment in question up to time t.

To evaluate the right-hand side of (5.2.15), we have to specify the initial condition. Let us assume that at $t = 0$ an even interval commences. Then the following mutually exclusive possibilities exist: either (a) m even and $(m - 1)$ odd intervals commence in $(0, t)$ and the time-point t is covered by the even interval, or (b) m even and m odd intervals commence in $(0, t)$ and the time-point t is covered by the odd interval. Adding the two contributions to the expected value and summing over m, we obtain

$$(5.2.16) \quad g(u, t)$$

$$= \sum_{m=1}^{\infty} \int_0^t \exp\left[\lambda_1(u - 1)x + \lambda_2(u - 1)(t - x)\right] [L_m(x)\beta_{m-1}(t - x)$$
$$+ \alpha_m(x) P_m(t - x)]\, dx,$$

where $\alpha_k(.)$ and $\beta_k(.)$ are the k-fold convolutions respectively of $\alpha(.)$ and $\beta(.)$, and $L_m(.)$ and $P_m(.)$ are given by

$$(5.2.17) \quad L_m(x) = \int_0^x \alpha_{k-1}(v)\,dv \int_{x-v}^{\infty} \alpha(w)\,dw$$

$$(5.2.18) \quad P_m(x) = \int_0^x \beta_{k-1}(v)\,dv \int_{x-v}^{\infty} \beta(w)\,dw.$$

The Laplace transform of $g(u, t)$ with respect to t has a convenient closed form:

$$(5.2.19) \quad g^*(u, s) = \int_0^{\infty} e^{-st} g(u, t)\,dt$$

$$= \frac{1}{1 - \alpha^*(s + \lambda_1 - \lambda_1 u)\beta^*(s + \lambda_2 - \lambda_2 u)}$$

$$\times \left[\frac{1 - \alpha^*(s + \lambda_1 - \lambda_1 u)}{s + \lambda_1 - \lambda_1 u} + \frac{\alpha^*(s + \lambda_1 - \lambda_1 u) - \alpha^*(s + \lambda_1 - \lambda_1 u)\beta^*(s + \lambda_2 - \lambda_2 u)}{s + \lambda_2 - \lambda_2 u} \right],$$

where $\alpha^*(.)$ and $\beta^*(.)$ are the Laplace transforms of $\alpha(.)$ and $\beta(.)$ respectively.

Returning to the problem in reliability analysis, we note that the expected time T to failure of the equipment can be obtained by simply setting $s = 0$, $u - 0$:

$$(5.2.20) \quad E\{T\} = \frac{1}{1 - \alpha^*(\lambda_1)\beta^*(\lambda_2)} \left[\frac{1 - \alpha^*(\lambda_1)}{\lambda_1} + \alpha^*(\lambda_1) \frac{1 - \beta^*(\lambda_2)}{\lambda_2} \right].$$

Higher moments of T can be evaluated, and these will depend on the derivatives of $\alpha^*(.)$ and $\beta^*(.)$ evaluated respectively at λ_1 and λ_2.

5.3 A model with continuous stochastic intensity

We now consider a doubly stochastic Poisson process whose intensity is the sum of squares of two stationary Gaussian variables. The direct motivation for such a study stems from physical considerations. There are a number of problems in the domain of classical noise theory that can be formulated in terms of such a model (see, for example, Karp and Clark (1970) and Srinivasan (1971)). Instead of dealing with the sums of squares of two real variables, it is better to deal with a complex Gaussian process, the intensity being identified with the modulus squared. Thus the intensity $\Lambda(.)$ can be written as

$$(5.3.1) \quad \Lambda(t) = |V(t)|^2 = V_1^2(t) + V_2^2(t),$$

where $V_1(t)$ and $V_2(t)$ are assumed to be independently and identically distributed with zero mean values. The stationary Gaussian process is completely characterized by the autocorrelation function $\Gamma(.,.)$ defined by

$$(5.3.2) \qquad\qquad \Gamma(t, t') = E[V(t) V^*(t')],$$

84

which satisfies the condition (see, for example, Loève (1963))

$$(5.3.3) \qquad \Gamma^*(t, t') = \Gamma(t', t)$$

or

$$\Gamma(t, t') = \gamma(|t - t'|) \, e^{i\omega(t-t')},$$

where for simplicity we have assumed ω to be a constant.

The generating function governing the counting process as defined by (5.1.2) can be written as

$$(5.3.4) \qquad g(u, T) = \mathrm{E}\left[\exp(u - 1) \int_0^T |V(t)|^2 \, dt\right].$$

It is more convenient to deal with $Q(s, T)$, where

$$(5.3.5) \qquad Q(s, T) = g(1 - s, T).$$

To evaluate the expectation value implied by (5.3.4), we seek a Loève (1963) expansion of the random variable $V(t)$ in terms of an orthogonal set of functions over the interval $(0, T)$:

$$(5.3.6) \qquad V(t) = \Sigma \, C_m \phi_m(t),$$

where the complex random coefficients satisfy the relation

$$(5.3.7) \qquad \mathrm{E}\left[C_m^* C_n\right] = \lambda_m \delta_{mn}$$

and the eigen functions ϕ_m corresponding to the eigenvalues λ_m satisfy the integral equation

$$(5.3.8) \qquad \int_0^T \phi_m(t) \, \Gamma(t, t') \, dt = \lambda_m \phi_m(t').$$

The Gaussian nature of the process is reflected by the complex Gaussian distribution of the coefficients $\{C_m\}$, the joint density function of which is given by

$$(5.3.9) \qquad \pi(\{C_m\}) = \prod_m \frac{1}{\pi \lambda_m} \exp\left(-|C_m|^2/\lambda_m\right).$$

Using (5.3.6) and (5.3.9), we can rewrite the generating function $Q(s, T)$ in terms of the eigenvalues λ_m:

$$(5.3.10) \qquad Q(s, T) = \prod_k (1 + s\lambda_k)^{-1}.$$

Thus the determination of the generating function reduces to that of evaluating the infinite product corresponding to the eigenvalues of the integral equation (5.3.8). We shall demonstrate the method of obtaining the generating function for the simple case when $\gamma(.)$ is exponential:

$$(5.3.11) \quad \Gamma(t, t') = \bar{\Lambda} \exp\left[-\beta|t - t'| + i\omega(t - t')\right],$$

where $\bar{\Lambda}$ is the stationary mean value of $\Lambda(t)$. Defining

$$(5.3.12)^{(*)} \qquad \Phi(t) = \phi(t)\, e^{i\omega t}$$

$$(5.3.13) \qquad \psi(p) = \int_0^T e^{-pt}\, \Phi(t)\, dt,$$

we obtain from (5.3.8)

$$(5.3.14) \quad \psi(p) = \frac{s\,[(\beta + p)\,\psi(\beta) + (\beta - p)\,\psi(-\beta)\, e^{-(\beta+p)T}]}{2\bar{\Lambda}\beta s - \lambda(\beta + p)(\beta - p)}.$$

We next observe that $\psi(.)$ is an entire function of its argument. However, the denominator of the right-hand side of (5.3.14) has two zeros corresponding to

$$(5.3.15) \qquad p = p_{1,2} = \pm \sqrt{\beta^2 - \frac{2\beta s}{\lambda}\bar{\Lambda}}.$$

Thus analyticity of $\psi(p)$ yields the conditions

$$(5.3.16) \quad (\beta + p_i)\,\psi(\beta) + (\beta - p_i)\,\psi(-\beta)\, e^{-(\beta + p_i)T} = 0, \quad i = 1, 2.$$

Eliminating $\psi(\beta)$ and $\psi(-\beta)$, we have

$$(5.3.17) \quad (\beta + p_1)^2\, e^{-(\beta - p_1)T} - (\beta - p_1)^2\, e^{-(\beta + p_1)T} = 0.$$

Thus the zeros of the function $P(\xi)$ ($\xi = 1/\lambda$), where

$$(5.3.18) \quad P(\xi) = (\beta^2 + \rho^2)\sinh \rho T + 2\beta\rho \cosh \rho T$$

$$(5.3.19) \qquad \rho = \sqrt{\beta^2 - 2\beta\bar{\Lambda}\xi},$$

are the eigenvalues of the integral equation (5.3.8). We next note that $F(\xi) = P(\xi)/2\beta\rho$ has the same zeros as $P(\xi)$ and that $\beta/2\bar{\Lambda}$ is not an eigenvalue of the integral equation (5.3.8). However, $F(\xi)$ is an integral function of order $\frac{1}{2}$, so that by the Hadamard–Weierstrass factorization theorem (see Hille (1962)) we have the canonical representation

$$(5.3.20) \qquad F(\xi) = F(0)\, \prod_k \left(1 - \frac{\xi}{\xi_k}\right),$$

where the ξ_k are the zeros of $F(\xi)$. Comparing (5.3.20) with (5.3.10), we find that

$$Q(s, t) = \frac{F(0)}{F(-s)}$$

$$(5.3.21) \qquad = e^{\beta T}\left[\cosh \rho_0 T + \frac{1}{2}\left(\frac{\rho_0}{\beta} + \frac{\beta}{\rho_0}\right)\sinh \rho_0 T\right]^{-1},$$

$$(5.3.22) \qquad \rho_0 = [\beta^2 + 2\beta\bar{\Lambda}s]^{1/2}.$$

$^{(*)}$ We drop the suffix k for simplicity of notation.

The solution for $Q(s, t)$ as given above was obtained by Bedard (1966) and Jakeman and Pike (1968) in connection with the problem of photoelectric counting. It is possible to solve the integral equation (5.3.8) for any general $\gamma(.)$. The analysis, due to Srinivasan and Sukavanam (1972), becomes exceedingly tedious, and in view of its complexity we have relegated the general solution to the appendix of this chapter.

The product densities of the events can be directly related to the correlation function $\Gamma(t, t')$. Using the notation of Chapter 4, we find that

$$(5.3.23) \quad h_1(t) = \mathrm{E}\,[|V(t)|^2] = \Sigma\,|C_n|^2 = \text{a constant} = \bar{\Lambda}\ (\text{say}),$$

$$h_2(t_1, t_2) = \mathrm{E}\,[|V(t_1)|^2\,|V(t_2)|^2]$$

$$(5.3.24) \qquad = \mathrm{E}\,[V^*(t_1)\,V(t_1)\,V^*(t_2)\,V(t_2)].$$

Using the stationary Gaussian nature of $V(t)$ and making use of (5.3.3), we find that

$$h_2(t_1, t_2) = h_1(t_1)h_1(t_2) + |\Gamma(t_1, t_2)|^2$$

$$(5.3.25) \qquad = \bar{\Lambda}^2 + [\gamma(|t_2 - t_1|)]^2.$$

Thus we note that $[\gamma(|t_2 - t_1|)]^2$ can be identified with the actual correlation function defined in problem 2.9, above. If we denote by $\hbar_m(t_1, t_2, \dots, t_m)$ the actual correlation function, then it is easy to show that

$$\hbar_m(t_1, t_2, \dots, t_m) = (m-1)!\,\Gamma(t_1, t_2)\,\Gamma(t_2, t_3) \dots \Gamma(t_{m-1}, t_m)\,\Gamma(t_m, t_1)$$

$$(5.3.26) \qquad = (m-1)!\,\gamma(|t_1 - t_2|)\gamma(|t_2 - t_3|) \dots \gamma(|t_{m-1} - t_m|)\gamma(|t_m - t_1|).$$

Thus the product-density generating functional (see problem 2.2, above) can be written as

$$(5.3.27) \quad L[\xi]$$

$$= \exp\left[\sum_{m=1}^{\infty} \int\int \dots \int \hbar_m(t_1, t_2, \dots, t_m)\,\xi(t_1)\,\xi(t_2) \dots \xi(t_m)\,dt_1\,dt_2 \dots dt_m\right].$$

If we confine our attention to the interval $(0, T)$ and consider the special case

$$(5.3.28) \qquad\qquad \gamma(|t_1 - t_2|) = \text{a constant} = \bar{\Lambda},$$

then we obtain from (5.3.26)

$$(5.3.29) \qquad\qquad L[\xi] = \left[1 - \bar{\Lambda}\int_0^T \xi(t)\,dt\right]^{-1}.$$

The above result was derived by Srinivasan and Vasudevan (1967b). Of course, when $\gamma(|t_1 - t_2|)$ is of the form (5.3.11), explicit integration implied by the right-hand side of (5.3.27) does not appear feasible. The generating function $g(u, T)$ as defined by (5.3.4) can be obtained from (5.3.29) by replacing $\xi(t)$ by $(u - 1)$:

$$(5.3.30) \qquad g(u, T) = [1 + \bar{\Lambda}T - \bar{\Lambda}uT]^{-1}.$$

The probability distribution of the number of counts in $(0, T)$ is given by

$$(5.3.31) \qquad p(n, T) = (1 + \bar{n})^{-1}(1 + \bar{n}^{-1})^{-n},$$

where

$$\bar{n} = E[N(0, T)] = \bar{\Lambda}T.$$

The correlation of the number of counts in different intervals can be obtained directly, using the product densities. If we consider two consecutive non-overlapping intervals $(0, x)$ (x, y) and assume that the second-order correlation is described by (5.3.11), we can use formula (5.3.6):

$$E[N(0, x) N(x, y)]$$

$$= \bar{\Lambda} \int_{x}^{x+y} \left[\bar{\Lambda}u + \frac{\bar{\Lambda}}{2\beta}(1 - e^{-2\beta u}) \right] du - \bar{\Lambda} \int_{0}^{y} \left\{ \bar{\Lambda}u + \frac{\bar{\Lambda}}{2\beta}(1 - e^{-2\beta u}) \right\} du$$

$$(5.3.32) \qquad = \bar{\Lambda}^2 xy + \frac{\bar{\Lambda}^2}{4\beta^2}(e^{-2\beta(x+y)} - e^{-2\beta x} - e^{-2\beta y} + 1).$$

Higher-order correlations can be calculated in a similar way.

5.4 Self-exciting point processes

So far we have considered stochastic point processes in which the "Poisson parameter" $\Lambda(t)$ is a stationary random process, so that we could imagine $\Lambda(t)$ being determined for all t without reference to the point process. However, there are a number of instances in which the process is self-exciting in the sense that $\Lambda(t)$ depends on the location of points at which events of the point process occur prior to t. In space-charge limited diodes (see Rowland (1937)) it happens that while a crude picture emerges on the basis of a Poisson model for the arrival of electrons at the anode, a finer description consists in imagining that the arrival of the electron at the anode changes the potential, the change in potential being transient in nature. However, the change in the potential produces a change in the arrival rate, albeit transient in nature. Specifically, Rowland assumed that if λ_0 is the arrival rate at $t = t_0$ and an electron has arrived in $(0, \Delta)$, then $\Lambda(t)$ $(t > t_0)$ until the next arrival is given by

$$(5.4.1) \qquad \Lambda(t) = \lambda_0 - b\, e^{-a(t-t_0)} \quad (t > t_0), \quad (a > 0, b > 0).$$

A similar model for Barkhausen noise was proposed by Srinivasan and Vasudevan (1966), with $b < 0$, to account for the increase in the rate of magnetization which is again transient in nature. All these cases can be studied by assuming that $\Lambda(t)$ satisfies the equation

$$(5.4.2) \qquad \Lambda(t) = \nu + \int_{-\infty}^{t} p(t - u)\, dN(u),$$

where the point process is specified by

$$
(5.4.3) \quad \begin{cases} \Pr\left[N(t+\Delta)-N(t)=1 \,|\, N(u): u\leqslant t\right] = \Lambda(t)\,\Delta \\ \Pr\left[N(t+\Delta)-N(t)>1 \,|\, N(u): u\leqslant t\right] = o(\Delta). \end{cases}
$$

Such a model was studied by Hawkes (1971). Generally $p(v)$ decays rapidly as v increases. In principle, $\Lambda(t)$ should remain positive, but there are physical problems in which $\Lambda(t)$ can become negative. Space-charge limited shot noise as described by (5.4.1) is an example of this kind. In such a case the problem is extremely difficult, since no events of the point process can materialize during the time-intervals in which $\Lambda(t)$ remains negative (see, for example, Srinivasan (1971)). The case when $p(.)$ is given by

$$
(5.4.4) \quad\quad p(v) = b\,e^{-av} \quad a > b > 0, v \geqslant 0
$$
$$
\quad\quad\quad\quad = 0 \quad\quad\quad v < 0
$$

is capable of explicit treatment. The product densities of the process $N(t)$ of different orders were obtained by Srinivasan and Vasudevan (1966). Hawkes (1971) has discussed a more general case.

If we assume stationarity[*] and positivity of the function $p(.)$, then we have

$$
(5.4.5) \quad\quad \bar\lambda = E\left[\Lambda(t)\right] = v + \bar\lambda \int_{-\infty}^{t} p(t-u)\,du
$$

or

$$
\bar\lambda = v \Big/ \left[1 - \int_{0}^{\infty} p(u)\,du\right]
$$

where it must be further imposed that $\int_{0}^{\infty} p(u)\,du < 1$.

The covariance density defined by

$$
(5.4.6) \quad \mu(\tau) = E\left[dN(t+\tau)\,dN(t)\right]/(dt)^2 - \bar\lambda^2
$$

can be written as

$$
\mu(\tau) = E\left[\frac{dN(t)}{dt}\left\{v + \int_{-\infty}^{t+\tau} p(t+\tau-u)\,dN(u)\right\}\right] - \bar\lambda^2
$$
$$
(5.4.7) \quad = \int_{-\infty}^{\tau} p(\tau-v)\,\mu^{(c)}(v)\,dv,
$$

where $\mu^{(c)}(.)$ is the complete covariance function defined by

$$
(5.4.8) \quad\quad\quad \mu^{(c)}(v) = \mu(v) + \bar\lambda\,\delta(v).
$$

---

[*] It can be proved (see Srinivasan and Vasudevan (1971) and problem 5.7, below) that the point process $N(t)$ is stationary when $p(.)$ is of the form (5.4.4). Extension to the case when $p(.)$ is a superposition of a finite number of terms of the type (5.4.4) may not be difficult.

Thus for $\tau > 0$, we have

(5.4.9) $\quad \mu(\tau) = \bar{\lambda} p(\tau) + \int\limits_{-\infty}^{\tau} p(\tau - v)\, \mu(v)\, dv.$

An explicit solution can be obtained for the special case when $p(.)$ is given by (5.4.4):

(5.4.10) $\quad \mu(\tau) = \dfrac{b\, \bar{\lambda}(2a - b)}{2(a - b)}\, e^{-(a-b)\tau} \quad \tau > 0$

With the help of $\mu(\tau)$, a spectral analysis of the point process can be carried out.

The stochastic process $\Lambda(t)$ can be studied directly when $p(.)$ is exponential of the form (5.4.4). In such a case we can start the process at $t = 0$ by assuming that it is switched on with a parametric value $\Lambda(0) = \nu$ with no memory. We can then seek the stationary solution of the problem. By our assumption, the sample function $\Lambda(t)$ is given by

(5.4.11) $\qquad\qquad \Lambda(t) = \nu + \sum_{i} b\, e^{-a(t-t_i)},$

where $t_1, t_2, \ldots$ are the points at which the events of the point process $N(t)$ occur. It is clear that the sample function decays exponentially at a rate a and that it undergoes upward jumps of magnitude b at the epochs corresponding to the events of the point process. We next define the function $\pi(\lambda, \nu, t)$ by

(5.4.12) $\quad \pi(\lambda, \nu, t)d\lambda = \Pr\{\lambda < \Lambda(t) < \lambda + d\lambda \mid \Lambda(0) = \nu\}$

and observe that the process $\Lambda(t)$ is quasi-Markovian[*] in the sense that the probability density function of $\Lambda(t + \Delta)$ can be expressed in terms of that of $\Lambda(t)$, provided the initial value $\Lambda(0) = \nu$ is known. Using this idea and equation (5.4.3), we obtain the Kolmogorov forward differential equation

(5.4.13) $\quad \dfrac{\partial \pi(\lambda, \nu, t)}{\partial t}$

$$= -a(\nu - \lambda)\dfrac{\partial \pi(\lambda, \nu, t)}{\partial \lambda} + (a - \lambda)\pi(\lambda, \nu, t) + (\lambda - b)\pi(\lambda - b, \nu, t) \quad t > 0,$$

satisfying the initial condition

(5.4.14) $\qquad\qquad \pi(\lambda, \nu, 0) = \delta(\lambda - \nu).$

It is interesting and worthwhile to note that the conditional probability density function $\pi^c(\lambda, \nu, t_2 \mid \lambda_1, t_1)$ defined by

---

[*] This terminology is due to Ramakrishnan (1955).

$$(5.4.15) \quad \pi^c(\lambda, \nu, t_2 | \lambda_1, t_1) d\lambda$$

$$= \Pr\{\lambda < \Lambda(t_2) < \lambda + d\lambda \mid \Lambda(0) = \nu, \Lambda(t_1) = \lambda_1\} \quad t_2 > t_1$$

for fixed λ_1 and ν satisfies (5.4.13) for $t_2 > t_1$, with the initial condition

$$(5.4.16) \qquad \pi(\lambda, \nu, t_1 | \lambda_1, t_1) = \delta(\lambda - \lambda_1).$$

Now the product densities of the process $N(t)$ are given by

$$(5.4.17) \qquad h_1(t) = E[\Lambda(t)]$$

$$(5.4.18) \quad h_2(t_1, t_2) = \lim_{\Delta \to 0} E[\Lambda(t_1)\Lambda(t_2) \mid N(t_1 + \Delta) > N(t_1)]$$

$$(5.4.19) \quad h_3(t_1, t_2, t_3)$$

$$= \lim_{\Delta_1, \Delta_2 \to 0} E[\Lambda(t_1)\Lambda(t_2)\Lambda(t_3) \mid N(t_i + \Delta_i) > N(t_i), i = 1, 2].$$

If we are in explicit possession of the right-hand side of (5.4.17) through (5.4.19), we can proceed to the limit as the origin recedes to minus infinity. If such a limit exists, we have then demonstrated the existence of the stationary point process (second- or third-order stationarity) $N(t)$, and we would incidentally have obtained the stationary product densities of the first few orders.

The mean value of $\Lambda(t)$ can be readily obtained from (5.4.13), although explicit solution of equation (5.4.13), is of itself not possible. We have

$$(5.4.20) \quad E[\Lambda(t)] = \frac{a\nu}{a-b} - \frac{b\nu e^{-(a-b)t}}{a-b},$$

so that

$$(5.4.21) \quad \lim_{t \to \infty} E[\Lambda(t)] = a\nu/(a-b),$$

in agreement with formula (5.4.5). The second-order product density can be evaluated by amplifying definition (5.4.18):

$$(5.4.22) \quad h_2(t_1, t_2) = \int E[\Lambda(t_2) | \Lambda(t_1 + 0) = \lambda + b, \Lambda(0) = \nu] \pi(\lambda, \nu, t_1) \lambda \, d\lambda.$$

The conditional expectation value of $\Lambda(t_2)$ as expressed above can be readily calculated by observing that $\pi^c(\lambda, \nu, t_2 | \lambda_1, t_1)$ as defined by (5.4.15) satisfies equation (5.4.13) for $t_2 > t_1$, with the initial condition (5.4.16). After some calculations (see problem 5.8, below) we obtain

$$(5.4.23) \quad E[\Lambda(t_2) | \Lambda(t_1 + 0) = \lambda + b, \Lambda(0) = \nu]$$

$$= a\nu[1 - e^{-(a-b)(t_2-t_1)}]/(a-b) + (\lambda + b) e^{-(a-b)(t_2-t_1)}.$$

Using (5.4.23) in (5.4.22), we obtain

(5.4.24) $h_2(t_1, t_2)$

$= \{av\,[1 - e^{-(a-b)(t_2-t_1)}]/(a - b) + b\,e^{-(a-b)(t_2-t_1)}\}\,E\,[\Lambda(t_1)]$

$+ e^{-(a-b)(t_2-t_1)}\,E\,\{[\Lambda(t_1)]^2\}.$

The second moment of $\Lambda(t_1)$ can easily be calculated from (5.4.13) (see problem 5.8, below):

(5.4.25) $E\,[\Lambda(t)]^2$

$= \{b^2\nu(2\nu - a + 2b)\,e^{-2t(a-b)} - 2b\nu(2a\nu + b^2)\,e^{-t(a-b)}$

$+ a\nu(2a\nu + b^2)\}/2(a - b)^2.$

Using (5.4.25) in (5.4.24) and proceeding to the limit as the origin recedes to minus infinity, we find that

(5.4.26) $\lim h_2(t_1, t_2) = (a\nu/a - b)^2 + a\nu b(2a - b)\,e^{-(a-b)\tau}/2(a - b)^2,$

where $\tau = |t_2 - t_1|$. The covariance density $\mu(\tau)$ can be obtained by observing that

(5.4.27) $\mu(\tau) = \lim \{h_2(t_1, t_2) - [h_1(t_1)]^2\}.$

On substituting from (5.4.26) and (5.4.21), we recover (5.4.10). Thus we have demonstrated that the point process $N(t)$ is stationary, to second order at least. In fact, stationarity of the point process can be demonstrated to any order by merely calculating the corresponding product density and establishing the existence of the limit as the origin recedes to minus infinity. It may be interesting and worthwhile to determine the largest class of functions $p(.)$ for which the stationary character of the point process $N(t)$ is ensured.

CHAPTER 5: PROBLEMS

5.1 Show that $g(u, T)$ as defined by (5.1.1) is given by

$$g(u, T) = \exp\{(\zeta - 1)\bar{\lambda}T + (\zeta - 1)\sigma_\lambda^2(e^{-\beta t} - 1 + \beta t)/\beta^2\},$$

where $\Lambda(t)$ is a stationary Uhlenbeck–Ornstein process with mean $\bar{\lambda}$ and variance σ_λ^2. (Cox and Lewis, 1966)

5.2 Discuss the reliability of a system which is failure free while not in operation and subject to a failure rate λ while in operation. If the system is switched on and off according to an alternative renewal process (see section **4.5**), with $\alpha(.)$ and $\beta(.)$ as the probability density functions governing the "on" and "off" periods, obtain the Laplace transform of the probability density function governing the time to failure. Deduce that the expected time to system-failure is given by

$$\frac{1}{\lambda} + \frac{\alpha^*(\lambda)}{1 - \alpha^*(\lambda)} \, \mathrm{E}\,[Q]$$

or

$$\frac{1}{\lambda} + \frac{\mathrm{E}\,[Q]}{1 - \alpha^*(\lambda)} \,,$$

where $\mathrm{E}\,[Q]$ is the expected duration of an "off" period and $\alpha^*(.)$ is the Laplace transform of $\alpha(.)$. (Gaver, 1963)

5.3 Obtain the characteristic functional of the counting process when the correlation is defined by (5.3.28). Deduce an explicit expression for the generating function of the multivariate distribution of the counts corresponding to several intervals (overlapping).

5.4 If the counting process is specified by the correlation function (5.3.28), show that $\omega_n(x)$, the density function governing the time $\tilde{T}_n$ to the nth event (measured from an arbitrary time-origin), is given by

$$\omega_n(x) \equiv n\bar{\Lambda}\,\frac{(\bar{\Lambda}x)^{n-1}}{(1 + \bar{\Lambda}x)^{n+1}}.$$

If, on the other hand, the counting process is specified by (5.3.11), the generating function of $\omega_n(x)$ is given by

$$H(u, x) = \sum_1^\infty u^n \omega_n(x)$$

$$= \frac{u}{u - 1} \frac{\partial}{\partial x} Q(1 - u, x).$$

In particular, show that

$$\omega_2(x) = \frac{e^{\beta x}\,\overline{\Lambda}\left[\cosh\rho_0 x - \dfrac{\beta}{\rho_0}\sinh\rho_0 x\right]}{\left[\cosh\rho_0 x + \dfrac{1}{2}\left(\dfrac{\rho_0}{\beta} + \dfrac{\beta}{\rho_0}\right)\sinh\rho_0 x\right]^2},$$

where $\rho_0 = \sqrt{\beta^2 + 2\beta\overline{\Lambda}}$.

5.5 Show that the analysis presented in section **5.3** leading to equation (5.3.29) can be extended to cover the case when $V(t)$ is a stationary Gaussian random process with non-zero and mean m, and that $L[\xi]$ is given by

$$L[\xi] = \frac{1}{1 - I_c \int_0^T \xi(t)\,dt}\,\exp\frac{I_s \int_0^T \xi(t)\,dt}{1 - I_c \int_0^T \xi(t)\,dt},$$

where $I_s = |m|^2$, $I_c = E[|V(t)|^2] - |m|^2$.

5.6 Solve the integral equation (5.4.9) when

$$p(v) = \sum_{j=1}^{k} \alpha_j\, e^{-\beta_j v} \quad (v > 0),$$

show that

$$\mu(\tau) = \sum_{j=1}^{k} \gamma_j\, e^{-\eta_j \tau} \quad (\tau > 0),$$

and identify the constants η_j. (Hawkes, 1971)

5.7 Examine whether the proof of stationarity of the self-exciting point process discussed in section **5.4** can be extended when

$$p(v) = av\, e^{-bv}.$$

5.8 If

$$p(n, v, t\,|\,\lambda_1, t_1) = \int_0^\infty \lambda^n \pi^c(\lambda, v, t\,|\,\lambda_1, t_1)\,d\lambda,$$

where π^c is defined by (5.4.15), show that

$$\frac{\partial p(n, v, t\,|\,\lambda_1, t_1)}{\partial t} = -na\, p(n, v, t\,|\,\lambda_1, t_1) + nav\, p(n-1, v, t\,|\,\lambda_1, t_1)$$

$$+ \sum_{i=1}^{n}\binom{n}{i} p(n-i+1, v, t\,|\,\lambda_1, t_1)\, b^i.$$

Setting $t_1 = 0$ and $\lambda_1 = v$, show that

$$E\{[\Lambda(t)]^2\}$$
$$= \{b^2 v(2v - a + 2b)\, e^{-2t(a-b)} - 2bv(2av + b^2)\, e^{-t(a-b)} + av(2av + b^2)\}/2(a-b)^2$$

$$E\{\Lambda(t)\} = (a - b\,e^{-t(a-b)})\nu/(a - b).$$

Using initial condition (5.4.16), show that for $t > t_i$,

$p(1, \nu, t \mid \lambda_1, t_1)$

$= a\nu\,[1 - \exp\{-(a - b)(t - t_1)\}]/(a - b) + \lambda_1 \exp\{-(a - b)(t - t_1)\}.$

5.9   Derive an explicit expression for the stationary product density of degree 3 of the counting process $N(t)$ described by (5.4.11).

5.10   Show that the general analysis presented in the Appendix to Chapter 5 holds good when

$$\Gamma(t, t') = f(|t - t'|) + i\,g(|t - t'|),$$

where $f(.)$ and $g(.)$ are respectively even and odd real-valued functions of their arguments, provided their Laplace transforms are rational functions vanishing at infinity faster than the reciprocal of the first power of the argument.   (Srinivasan and Sukavanam, 1972)

CHAPTER 5: APPENDIX

We shall here demonstrate the possibility of obtaining $Q(s, T)$ as defined by (5.3.5) explicitly for any arbitrary $\Gamma(.)$ of the form

$$(5.A.1) \qquad \Gamma(t, t') = f(|t - t'|) \exp[i\omega_0(t - t')],$$

where $f(.)$ is a real-valued function defined on the positive real axis. We find that $\phi(t)$ as defined by (5.3.12) satisfies the equation

$$(5.A.2) \qquad \int_0^T f(|t - t'|)\phi(t)\, dt = \lambda\phi(t').$$

To solve the above integral equation, we introduce the Laplace transform of $\phi(.)$ by

$$(5.A.3) \qquad \psi(p) = \int_0^T \phi(t) \exp(-pt)\, dt.$$

We shall assume that $f(.)$ admits of a Laplace transform $f^*(.)$. Feeding into equation (5.A.2) the inversion formula

$$(5.A.4) \quad f(|t - t'|) = [1/2\pi i] \int_{\sigma - i\infty}^{\sigma + i\infty} f^*(z) \exp(z|t - t'|)\, dz,$$

where σ is real and positive, we obtain

$$(5.A.5) \quad [1/2\pi i] \int_{\sigma - i\infty}^{\sigma + i\infty} f^*(z)\, dz\, \{[\psi(z) \exp[(z - p)T] - \psi(p)]$$

$$\div [z - p] + [\psi(-z) - \psi(p)]/[z + p]\}$$

$$= \lambda\psi(p).$$

We note that $f^*(z)$ is analytic in the half-plane $\operatorname{Re} z > 0$ by virtue of the bounded nature of the autocorrelation function as defined by (5.A.1). To make further progress we shall assume that $f^*(z)$ is a rational function of z:

$$(5.A.6) \qquad f^*(z) = g(z)/h(z),$$

where $h(z)$ is a polynomial of degree n and $g(z)$ a polynomial of degree not exceeding $(n - 1)$. It may not be difficult to extend these arguments to a more general case where $f^*(z)$ admits of a Mittag–Leffler expansion. We evaluate the line integral occurring on the left-hand side of (5.A.5) by choosing σ to be greater than max $(0, \operatorname{Re} p, -\operatorname{Re} p)$. We observe that $\psi(.)$ is an entire function of its argument, and using the definition we obtain the following estimate for large $\operatorname{Re} p$:

$$(5.A.7) \quad |\psi(p)| \simeq [1 - \exp(-\operatorname{Re} p.T)]/\operatorname{Re} p \quad |\operatorname{Re} p| \gg 1.$$

Using the estimate of (5.A.7), we evaluate the line integral corresponding to the second and fourth terms of the integrand by closing to the right and conclude that it is zero.

Next, we observe that the poles of $f^*(z)$ are due to the zeros of $h(z)$ which are necessarily located in the half-plane Re $z < 0$. Assuming that $h(z)$ has m distinct zeros at the points $z_1, z_2, \ldots, z_m$ with respective multiplicities $l_1, l_2, l_3, \ldots, l_m$, we evaluate the line integrals corresponding to the first and third terms on the left-hand side of (5.A.5) by converting them into contour integrals. We thus obtain

$$(5.A.8) \quad \sum_{k=1}^{m} [a_k \exp(-pT) + b_k] + [f^*(p) + f^*(-p)]\psi(p) = \lambda\psi(p),$$

where

$$(5.A.9a) \qquad\qquad a_k = \sum_{j=0}^{l_k-1} F_{jk}/(p - z_k)^{j+1},$$

$$(5.A.9b) \quad F_{jk} = \frac{1}{(l_k - 1 - j)!} \left(\frac{d}{dz}\right)^{l_k-1-j} [e^{zT}\psi(z)f^*(z)],$$

$$(5.A.9c) \qquad\qquad b_k = \sum_{j=0}^{l_k-1} G_{jk}/(p + z_k)^{j+1},$$

$$(5.A.9d) \quad G_{jk} = \frac{(-)^j}{(l_k - 1 - j)!} \left(\frac{d}{dz}\right)^{l_k-1-j} [\psi(-z)f^*(z)].$$

The terms $\Sigma\, a_k$ and $\Sigma\, b_k$ can be cast into the form

$$(5.A.10a) \quad \Sigma\, a_k = \frac{X(p)}{h(p)} \equiv \sum_{i=0}^{n-1} \frac{X_i p^i}{h(p)},$$

$$(5.A.10b) \quad \Sigma\, b_k = \frac{Y(p)}{h(-p)} \equiv \sum_{i=0}^{n-1} \frac{Y_i p^i}{h(-p)},$$

where the constants $\{X_i\}$ and $\{Y_i\}$ are linear combinations of $\{F_{jk}\}$ and $\{G_{jk}\}$. Thus (5.A.8) can be written in the form

$$(5.A.11) \quad \psi(p) = \frac{h(-p)X(p)\,e^{-pT} + h(p)Y(p)}{\lambda h(-p)h(p) - g(p)h(-p) - g(-p)h(p)}.$$

The denominator of the right-hand side of (5.A.11) is an even polynomial of degree $2n$. If $\{+p_i\}\, i = 1, 2, \ldots, 2n$ are its zeros, then the set $\{+p_i\}$ can be regarded as the $2n$-branches of a multiple-valued function of λ defined by

$$(5.A.12) \quad \lambda h(p)h(-p) - g(p)h(-p) - g(-p)h(p) = 0.$$

By virtue of the even nature of the polynomial, the set can without loss of generality be denoted by $\{\pm p_i, i = 1, 2, \ldots, n\}$. We next observe that by virtue of $\psi(.)$ being an entire function of its argument, the numerator of the right-hand side of (5.A.11) vanishes at $\pm p_i$:

$$(5.A.13) \quad \begin{cases} h(-p_i)\,X(p_i)\,e^{-p_iT} + h(p_i)\,Y(p_i) = 0 \\ h(p_i)\,X(-p_i)\,e^{p_iT} + h(-p_i)\,Y(-p_i) = 0. \end{cases}$$

Equations (5.A.13), which are $2n$ in number, determine the constants $\{X_i\}$ and $\{Y_i\}$. Equations (5.A.12) can be written in the form

$$(5.A.14) \qquad\qquad \mathbf{D}\,\vec{\mathbf{K}} = 0,$$

where $\vec{\mathbf{K}}$ is the column vector with elements $X_0, X_1, \ldots, X_{n-1}, Y_0, Y_1, \ldots, Y_{n-1}$ and $\mathbf{D}$ is a $2n \times 2n$ matrix with elements D_{ij} given by

$$(5.A.15) \qquad D_{ij} = \begin{cases} h(-p_i)\,e^{-p_iT}\,p_i^j & j \leqslant n, i \leqslant n \\[4pt] h(p_i)\,p_i^{j-n} & j > n, i \leqslant n \\[4pt] h(p_{i-n})\,e^{p_{i-n}T}\,(-p_{i-n})^j & j \leqslant n, i > n \\[4pt] h(-p_{i-n})\,(-p_{i-n})^{j-n} & j > n, i > n. \end{cases}$$

We notice that (5.A.14) represents a homogeneous system of equations for the components of $\vec{\mathbf{K}}$, and in order that the solution be non-trivial, we have

$$(5.A.16) \qquad\qquad |\mathbf{D}| = 0.$$

Since λ is the only unknown parameter in the determinant, the above equation is the eigenvalue equation for λ. Since the p_i's are functions of λ defined by (5.A.12), it is clear that there are an infinite number of eigenvalues. If we denote $|\mathbf{D}|$ by $F(1/\lambda)$, it is easy to see that $F(.)$ is not an entire function of its argument in view of its not returning to its original value if we go along any arbitrary closed contour containing the origin. However, it can easily be proved that $F(.)$ is an analytic function in any bounded domain of the cut ξ-plane ($\xi = 1/\lambda$). Thus the non-analytic property in the ξ-plane arises essentially from the multiple-valued nature of the function defined by (5.A.15). But it is easily seen that the determinant is an alternant with reference to the parameters $p_1, p_2, \ldots, p_{2n}$, and that the different values of $|D(\xi)|$ corresponding to the different branches are obtained by permutations of the parameters $p_1, p_2, \ldots, p_{2n}$. By the characteristic property of the determinant, all the values are the same except for the sign. Thus we can construct an entire function from $|D(\xi)|$ by multiplying or dividing it by an appropriate alternant in $p_1, p_2, \ldots, p_{2n}$.

Defining $P(\xi)$ by

$$(5.A.17) \qquad\qquad P(\xi) = |D(\xi)| / |D_0(\xi)|,$$

where the elements of the determinant $|D_0(\xi)|$ are given by

$$(5.A.18) \quad |D_0(\xi)|_{ij} = p_i^{j-1} \quad i, j = 1, 2, \ldots, 2n,$$

we note that $|D_0(\xi)|$ is a factor of $|D(\xi)|$ and that the zeros of $|D_0(\xi)|$ do not contribute to the eigenvalues of the original integral equation (5.A.2). Now $P(\xi)$ is an entire function and its zeros are directly related to the eigenvalues of the basic integral equation (5.A.2). We can obtain a representation for $P(\xi)$ by using the Hadamard–Weierstrass theorem relating to the canonical representation of an entire function (see, for example, Einar Hille (1962)). To do this, we first determine the order of $P(\xi)$. We notice that the order of $P(\xi)$ is related to the behaviour of the p_i's for large values of ξ. If the degree of $g(p)$ is m (which is always less than n) it follows from (5.A.12) that for large $|\xi|$

(5.A.19) $\quad |p| \simeq C|\xi|^{1/2(n-m)}$, C a positive constant.

Observing that the dependence of $P(\xi)$ on ξ, which in turn is related to the dependence of $|D(\xi)|$ on ξ, is dominated by terms of the form $\exp(-p_iT)$, we conclude that the order of the entire function $P(\xi)$ is at most $\frac{1}{2}$.

Thus we have the representation for $P(\xi)$:

(5.A.20) $\qquad\qquad P(\xi) = P(0) \prod_k (1 - \xi/\xi_k),$

where the constant $P(0)$ is determined using (5.A.17). Comparing the above representation with (5.3.10), we obtain the following explicit expression for the generating function:

(5.A.21) $\qquad\qquad Q(s, T) = P(0)/P(-s).$

The results corresponding to the special case dealt with in section **4.3** can be easily deduced from (5.A.21). In this case $|D(\xi)|$ is of order 2, given by

(5.A.22) $\quad \begin{vmatrix} (\beta - p_1)\exp[-(\beta + p_1)T] & (\beta + p_1) \\ (\beta - p_2)\exp[-(\beta + p_2)T] & (\beta + p_2) \end{vmatrix}$

where $p_1 = -p_2 = [\beta^2 - 2\beta\bar{\Lambda}\xi]^{1/2}$. The corresponding alternant $|D_0(\xi)|$ in this case cannot be other than $p_2 - p_1$. Thus $P(\xi)$ is given by

(5.A.23) $\quad -\frac{1}{2}[\beta^2 - 2\beta\bar{\Lambda}\xi]^{1/2} \begin{vmatrix} (\beta - p_1)\exp[-(\beta + p_1)T] & (\beta + p_1) \\ (\beta - p_2)\exp[-(\beta + p_2)T] & (\beta + p_2) \end{vmatrix}$

from which we arrive at the generating function

(5.A.24) $\quad Q(s, T) = e^{\beta T}[\cosh \rho_0 T + \frac{1}{2}(\rho_0/\beta + \beta/\rho_0)\sinh \rho_0 T]^{-1}$

with $\quad \rho_0 = \sqrt{\beta^2 + 2\beta\bar{\Lambda}s}$.

6   MULTIVARIATE POINT PROCESSES

6.1   Introduction

In the previous chapters we have dealt with the theory of point processes in which the events are all of the same type, the only characteristic by which different events are distinguishable being the position of the event in the one-dimensional continuum. But there also arise a number of situations in which events of two or more types occur in the one-dimensional continuum. In such a case the point process may be visualized as a superposition of as many point processes as the types of events. Superposition of point processes was first studied explicitly by Palm (1943) in an attempt to explain the Poisson nature of telephone calls. A rigorous proof in the form of a limit theorem was provided later by Khintchine (1955). An extension to m-dimensional point processes has recently been obtained by Cinlar (1968). However, these studies relate to the asymptotic behaviour of the superposition of a large number of point processes and do not throw any light on problems that arise from the superposition of two- or three-point processes. In this chapter we shall discuss such superpositions and attempt to characterize the resulting point process by extensions of the methods developed in Chapters 2 and 4.

Superpositions of point processes or of series of events arise in many contexts. The following examples will serve to bring out the idea.

(i)   In industrial processes in which dissimilar machines in series or in parallel operate, the epochs of breakdown of the machines constitute a superposition of point processes, each point process representing the failure pattern of any particular machine.

(ii)   In queuing theory (particularly in queues with one server) the arrivals and departures constitute two different point processes. In a general theory, each point process depends on the other, thus introducing an inherent interaction between the two. Studies relating to the joint properties of arrival and departure patterns are of very recent origin (see Daley (1968) and Srinivasan et al. (1971)). It is expected that many fruitful general results characterizing the queuing process will be obtained in this manner (see, for example, D.G. Kendall (1964)).

(iii) Studies in neurophysiology by Bishop, Levick and Williams (1964) appear to indicate that the information-theoretic aspects of neuronal firings (responses) can be best studied by a model corresponding to the superposition of excitatory and inhibitory events. Several models dealing with special types of dependence of the responses on the excitatory and inhibitory mechanism have been proposed (see, for example, ten Hoopen and Reuver (1965, 1967a); Srinivasan and Rajamannar (1970a, b, c); and Srinivasan, Rajamannar and Rangan (1971). Here again it is expected that general results can be derived from the particular dependency structure of the interaction (see Lawrance (1970, 1971)).

(iv) An example of a non-stationary branching bivariate process is provided by electromagnetic cascades (see Bhabha and Heitler (1937) and Bhabha (1950)). A bivariate point-process formulation results if we consider the process in the Cartesian product space of energy and thickness of material traversed by the electrons and photons (Srinivasan (1969)).

(v) Problems of a nature very similar to electromagnetic cascades arise in population growth if we take into account the different species. The points here may be regarded either as birth or as extinction.

The point processes described above can also be regarded as a univariate point process in which a real-valued quantity is associated with each point event. An example of this is provided by electromagnetic showers in which the point events correspond to the locations at which electron—positron pairs are formed (see Ramakrishnan and Srinivasan (1956) and Srinivasan (1969)). If, however, we specialize to the case when the real-valued quantity is capable of assuming only two possible values, we realize a bivariate point process.

Bivariate point processes arising from the superposition of two independent renewal processes were studied by McFadden and Weissblum (1963). Recently the properties of a pair of interacting renewal processes with special reference to the marginal distribution of the inhibited process (where the interaction is characterized by the inhibition of one of the processes by the other) have been studied by Srinivasan and Rajamannar (1970a, b). However, a systematic analysis of bivariate point processes — a study which proceeds almost parallel with the development of the theory of univariate stationary processes as presented in Chapter 4 — is essentially due to Cox and Lewis (1970). From a general point of view, a bivariate or a multivariate point process can be regarded as a special point process, and as such the characterization of the process on the lines of Chapter 4 should be adequate. But the added information in the form of the special properties of the univariate point processes (arising from the different types of events that constitute the multivariate point process) helps in the characterization of the point process in terms of intensity and correlation functions. Accordingly we shall attempt a parallel development of the theory of bivariate point processes on lines similar to those of Chapter 4.

6.2 General properties of bivariate processes

Let us consider a bivariate point process of events of two types, called type a and type b. Most of our arguments are equally applicable to any multivariate process. The point process generated by events of only one type (a or b) is called a *marginal process*. We can first extend the notion of "regularity". The bivariate process is said to be *regular* if the superposed point process (regardless of the type) is regular, and *marginally regular* if the marginal point processes are regular. Thus, "regularity" implies marginal regularity; however, a marginally regular point process need not be regular (see problem 6.1, below).

Just as in the case of univariate point processes, we can introduce three types of stationarity: simple stationarity, weak stationarity, and complete stationarity. Let $N_a(t, x)$ $(N_b(t, x))$ denote the number of events of type a (b) in the interval $(t, t + x)$. A bivariate point process is said to be

(i) "simply stationary" if

(6.2.1) $\quad \Pr\left[N_a(t, x) = n, N_b(t', x') = n'\right]$

$\quad\quad = \Pr\left[N_a(t + h, x) = n, N_b(t' + h, x') = n'\right]$

for all $t, x, h > 0, n \geqslant 0$ and $n' \geqslant 0$;

(ii) "weakly stationary" if

(6.2.2) $\quad \Pr\left[N_a(t_1, x_1) = n_1, N_b(t'_1, x'_1) = n'_1, N_a(t_2, x_2) = n_2, N_b(t'_2, x'_2) = n'_2\right]$

$\quad\quad = \Pr\left[N_a(t_1 + h, x_1) = n_1, N_b(t'_1 + h, x'_1) = n'_1,\right.$

$\quad\quad\quad \left. N_a(t_2 + h, x_2) = n_2, N_b(t'_2 + h, x'_2) = n'_2\right]$

for all $t_1, t_2, t'_1, t'_2, x_1, x_2, x'_1, x'_2, h > 0, n_1 \geqslant 0, n_2 \geqslant 0, n'_1 \geqslant 0$ and $n'_2 \geqslant 0$;

(iii) "completely stationary" if

(6.2.3) $\quad \Pr\left[N_a(t_i, x_i) = n_i, N_b(t_j, x_j) = n'_j, \quad i = 1, 2, \ldots, m, \ j = 1, 2, \ldots, m'\right]$

$\quad\quad = \Pr\left[N_a(t_i + h, x_i) = n_i, N_b(t_j + h, x_j) = n'_j,\right.$

$\quad\quad\quad \left. i = 1, 2, \ldots, m, \ j = 1, 2, \ldots, m'\right]$

for all $t_i, x_i, t_j, x_j, n_i \geqslant 0 \ (i = 1, 2, \ldots, m), n'_j \geqslant 0 \ (j = 1, 2, \ldots, m'), h > 0$, and for all positive integral values of m and m'. In what follows we shall assume the process to be completely stationary unless otherwise specified.

Just as in the case of univariate point processes, we can characterize the process by one or other of the following random variables:

(i) $\quad N(t, x) = (N_a(t, x), N_b(t, x))$,

(ii) $\quad L_n(t, \alpha)$, the time from an arbitrary point t to the nth event of type α $(\alpha = a, b)$,

(iii) $\quad L_n^\alpha(t, \beta)$, the time from an arbitrary event of type α at t to the nth event of type β.

We note that the random variables $L_n(t, \alpha)$ and $L_n^\alpha(t, \beta)$ $(\alpha \neq \beta)$ are the corresponding generalizations of forward recurrence times of the univariate point process, and by allowing n to take negative integral values we obtain the corresponding backward recurrence times.

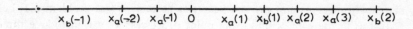

Fig. 6.1   The events corresponding to a bivariate process plotted on the time axis

We can mark the points corresponding to type a and type b events on the t-axis with reference to an arbitrary origin by the points $x_a(1), x_a(2),$ $\ldots, x_a(n), \ldots, x_b(1), x_b(2), \ldots, x_b(n), \ldots$, (see Fig. 6.1). We shall use the symbol y to denote the intervals, so that

$$(6.2.4) \qquad y_\alpha(i) = x_\alpha(i+1) - x_\alpha(i), \quad \alpha = a, b$$

represents the interval length between the $(i+1)$th and the ith events of type α, counted from the origin. If we do not wish to distinguish between the two types of events, we shall simply use the symbols x_i and y_i so as to be consistent with the notation of section **4.1**. It is convenient to introduce the following density functions:

$$\Pr\left[x < L_n(t, \alpha) < x + dx\right] = \Pr\left[x < L_n(0, \alpha) < x + dx\right]$$

$$(6.2.5) \qquad\qquad\qquad = g_n(x, \alpha)\, dx \quad (\alpha = a, b);$$

$$\Pr\left[x < L_n^\alpha(t, \beta) < x + dx\right] = \Pr\left[x < L_n^\alpha(0, \beta) < x + dx\right]^{(*)}$$

$$(6.2.6) \qquad\qquad\qquad = g_n^\alpha(x, \beta)\, dx \quad (\beta = a, b);$$

$$(6.2.7) \qquad f_n(x, \alpha)\, dx = \Pr\left[x < S_\alpha(n) < x + dx\right],$$

where

$$S_\alpha(n) = \sum_{i=k}^{n+k-1} (x_\alpha(i+1) - x_\alpha(i))$$

$$(6.2.8) \qquad\qquad = \sum_{i=k}^{n+k-1} y_\alpha(i) \quad (k \text{ an arbitrary integer});$$

$$P_\alpha(n, x) = \Pr\left[N_\alpha(t, x) = n\right]$$

$$(6.2.9) \qquad\qquad = \Pr\left[N_\alpha(0, x) = n\right].$$

As in Chapter 4, we shall use capital letters to denote the distribution functions. Apart from these distribution functions characterizing marginal point

(*) As in Chapter 4, we are assuming that complete stationarity with reference to the multivariate distribution of $N(t, x)$ implies stationarity with reference to the multivariate distribution of $L_n(t, \alpha)$ (see problem 4.2).

processes, we can also introduce joint density functions of $L_n(t, a)$ and $L_n(t, b)$ and perhaps of any two members of $L_n^\alpha(t, \beta)$. The identities established in section **4.2** are valid in the present context also. The generalized product densities are defined for the marginal processes, and it may be worthwhile to characterize the process through these functions.

The superposed process without specification of the type of event is called a *pooled process*. Perhaps the best way to describe the pooled process is by means of the joint characteristic functional $\Phi(\theta, t)$, defined by[*]

$$\Phi(\theta, t) = E\{\exp i \int \theta(u) d_u N.(t, u)\}$$

(6.2.10)
$$= E\{\exp i \int \theta(u) [d_u N_a(t, u) + d_u N_b(t, u)]\}.$$

The characteristic functional of any process is a fairly difficult choice; usually the first few functional derivatives, which are simply the generalized product densities of the pooled process, may be convenient objects to deal with. In any particular problem they are obtained directly by incorporating the mutual dependence (or independence, if that be the case) of the constituent marginal processes. We shall illustrate this point in the next section, when we deal with dependence and correlation.

So far we have attempted to describe the bivariate process in terms of the distribution functions of the marginal processes. However, there are certain aspects of the problem which are quite characteristic of bivariate processes. This can be brought out vividly by attempting analogues of the Palm–Khintchine formula connecting the joint distribution of the pair $L_n(t, \alpha)$[†] and $S.(n)$. To illustrate the point, let us consider the joint distribution of $L_1(a)$ and $L_1(b)$. Let

(6.2.11) $\quad g_1^{ab}(x, y) \, dx \, dy = \Pr [x < L_1(a) < x + dx, y < L_1(b) < y + dy]$

(6.2.12) $\quad \begin{cases} h_1^\alpha(x) \, dx = E[d_x N_\alpha(t, x)] \\ \qquad = E[d_x N_\alpha(0, x)] \quad (\alpha = a, b). \end{cases}$

By virtue of stationarity of the pooled process, it follows that $h_1^\alpha(.)$ is a constant equal to μ_α, the rate of events of type α. If the process is in addition regular, these rates can be identified with the stationary probabilities (per unit of the parameter) of occurrence of one or more events (Korolyuk's theorem). If

---

[*] We shall use the symbol "." in place of the type of event whenever we deal with the pooled process.

[†] Since we deal with completely stationary processes, we shall use the symbol $L_n(\alpha)$ to denote $L_n(0, \alpha)$. $S.(n)$ denotes the sum corresponding to (6.2.8) of the pooled process.

104

(6.2.13) $fg_{11}^{\alpha\beta}(x, y)\, dx\, dy = \Pr[x < y_\alpha(1) < x + dx, y < L_1^\alpha(\beta) < y + dy]$

$$\alpha \neq \beta,$$

we find by using arguments similar to the derivation of equation (4.2.1) that

(6.2.14) $g_1^{ab}(x, y)$

$$= \mu_a \int_0^\infty fg_{11}^{ab}(x + u, y + u)\, du + \mu_b \int_0^\infty fg_{11}^{ba}(y + u, x + u)\, du.$$

By definition of $g_1^{ab}(.,.)$ it follows that

(6.2.15a) $$g_1^a(x) = \int_0^\infty g_1^{ab}(x, y)\, dy$$

(6.2.15b) $$g_1^b(y) = \int_0^\infty g_1^{ab}(x, y)\, dx.$$

Multiplying (6.2.15a) and (6.2.15b) by x and y respectively and integrating, we obtain

(6.2.16) $\mathrm{E}[L_1(0, a)] + \mathrm{E}[L_1(0, b)]$

$$= \tfrac{1}{2}\mu_a \mathrm{E}[y_a^2(1)] + \tfrac{1}{2}\mu_b \mathrm{E}[y_b^2(1)].$$

Multiplying both sides of (6.2.14) by $x + y$ and integrating over x and y, we also obtain

$$\mathrm{E}[L_1(0, a)] + \mathrm{E}[L_1(0, b)]$$

$$= \mu_a \mathrm{E}[y_a(1)L_1^a(b)] + \mu_b \mathrm{E}[y_b(1)L_1^b(a)]$$

(6.2.17) $$= \sum_\alpha \mu_\alpha \mathrm{E}[y_\alpha(1)L_1^\alpha(\beta)] \quad \beta \neq \alpha.$$

Relations of the type (6.2.16) and (6.2.17) appear to have been noticed by Wisniewski (see Cox and Lewis (1970)). It is possible to obtain relations connecting the joint distribution function of $L_n(\alpha)$ with that of $S_n(\alpha)$ and $L_n^\alpha(\beta)$ (see problem 6.2, below).

We next observe that the density function $f(.)$ governing any two successive events of the pooled process can be expressed in terms of the joint density functions of the bivariate process. By virtue of stationarity, if follows that

$$f(x) = \mu_a/(\mu_a + \mu_b) \int_x^\infty [fg_{11}^{ab}(x, y) + fg_{11}^{ab}(y, x)]\, dy$$

(6.2.18) $$+ \mu_b/(\mu_a + \mu_b) \int_x^\infty [fg_{11}^{ba}(x, y) + fg_{11}^{ba}(y, x)]\, dy.$$

6.3 Intensity functions and generalized product densities

We now proceed to discuss the correlational structure of bivariate point processes. The first-order properties are best described by means of the intensity functions introduced in section **4.3**. If the process is marginally regular, then there are three intensity functions defined by

$$\lambda_\alpha(t|\mathcal{H}_t) = \lim_{\Delta \to 0} \Pr[N_\alpha(t, \Delta) \geq 1|\mathcal{H}_t]/\Delta$$

$$(6.3.1) \qquad = \Pr[d_t N_\alpha(0, t) \geq 1|\mathcal{H}_t]/dt \quad \alpha = a, b$$

$$\lambda_{ab}(t|\mathcal{H}_t) = \lim_{\Delta \to 0} \Pr[N_a(t, \Delta)N_b(t, \Delta) \geq 1|\mathcal{H}_t]/\Delta$$

$$(6.3.2) \qquad = \Pr[d_t N_a(0, t) d_t N_b(0, t) \geq 1|\mathcal{H}_t]/dt.$$

If the process is regular, then $\lambda_{ab}(t|\mathcal{H}_t) = 0$ and μ_α, the rate of the events of type α, can be identified with the stationary value of λ_α. If the process is completely stationary (but not necessarily regular), then the intensity functions depend only on $\mathcal{H}_t$ and not on t. For a general bivariate process which is not necessarily marginally regular, a characterization is provided by the intensity functions

$$(6.3.3) \quad \mu_\alpha(t, \mathcal{H}_t) = \lim_{\Delta \to 0} E[N_\alpha(t, \Delta)|\mathcal{H}_t]/\Delta$$

$$= E[d_t N_\alpha(0, t)|\mathcal{H}_t]/dt$$

$$(6.3.4) \quad \mu_{ab}(t, \mathcal{H}_t) = E[d_t N_a(0, t) d_t N_b(0, t)|\mathcal{H}_t]/dt.$$

Thus for a stationary bivariate process we have the following analogues of (4.3.8):

$$(6.3.5) \quad \lambda_\alpha(0|\mathcal{H}_0) \leq \sum_i i\lambda_\alpha^i(0|\mathcal{H}_0) = \mu_\alpha(0|\mathcal{H}_0),$$

where

$$(6.3.6) \quad \lambda_\alpha^i(0|\mathcal{H}_0) = \lim_{\Delta \to 0} \Pr[N_\alpha(0, \Delta) = i|\mathcal{H}_0]/\Delta,$$

so that $\lambda_\alpha(0|\mathcal{H}_0) = \mu_\alpha(0|\mathcal{H}_0)$ if and only if

$$(6.3.7) \quad \Pr[N(0, \Delta) = i|\mathcal{H}_0] = o(\Delta) \quad i > 1,$$

Korolyuk's criterion for marginal regularity.

It also follows by the ordinary renewal theorem of Chapter 3 that $\mu_a + \mu_b$ should be equal to the reciprocal of mean value of the interval length between any two successive events of the superposed process. Using relation (6.2.18), we arrive at the following identity:

$$(6.3.8) \quad \sum_{\substack{\alpha, \beta \\ \alpha \neq \beta}} \mu_\alpha \int_0^\infty x \, dx \int_x^\infty [fg_{11}^{\alpha\beta}(x, y) + fg_{11}^{\alpha\beta}(y, x)] \, dy = 1.$$

To illustrate how the intensity specification can be achieved in any particular process, let us consider the two-state semi-Markov process introduced in Chapter 4. Here $\mathcal{H}_t$ can be restricted to be the backward recurrence time to the previous event and the type of that event, which can be denoted by the pair $\{L_{-1}(.), .\}$. Thus we can write the partial history (relevant to the process) as (u, α). If the process is regular, we have

$$(6.3.9) \quad \lambda_\alpha(t \mid (u, \beta)) = \frac{p_{\beta\alpha} \, g_1^\beta(u, \alpha)}{1 - p_{\beta a} \, G_1^\beta(u, a) - p_{\beta b} \, G_1^\beta(u, b)} \,,$$

where $g_1^\beta(u, \alpha)$ (or $G_1^\beta(u, \alpha)$) is to be interpreted as $f_\alpha(u)$ (or $F_\alpha(u)$) whenever $\alpha = \beta$.

We observe that the intensity specification of a stationary process is unique in the sense that two different intensity specifications will always lead to different intensities of events for at least one history of non-zero probability. It is also easy to see that events of types a and b are independent if and only if $\lambda_a(t \mid \mathcal{H}_t)$ and $\lambda_b(t \mid \mathcal{H}_t)$ are respectively independent of the projected histories $\mathcal{H}_t^{(b)}$ and $\mathcal{H}_t^{(a)}$.

There are some special processes in which both the quantities $\lambda_a(t \mid \mathcal{H}_t)$ and $\lambda_b(t \mid \mathcal{H}_t)$ depend on $\mathcal{H}_t$ only through the projection $\mathcal{H}_t^{(a)}$. Cox and Lewis (1970) prefer to call such processes "purely a-dependent" processes. An example of such a process is provided by a model of neuronal spike trains proposed by ten Hoopen and Reuver (1968). The model incorporates selective interaction between two dependent, recurrent time-sequences of stimuli, called "excitatory" and "inhibitory". The excitatory process, hereinafter called the "a-process", is assumed to be a stationary renewal point process. Each a-event triggers, in a manner to be specified presently, an inhibitory point process called the "i-process", the sequence of i-events originating from a particular a-event continuing until the next a-event, the a-event itself being deleted by this sequence of i-events. Every undeleted a-event is assumed to give rise to a response (called "b-event"). The quantity of interest from the neurophysiological point of view is the sequence of b-events. There are two models of interaction, called model I and model II, according to whether a-events or b-events give rise to the sequence of inhibitories. We can visualize the process as a bivariate point process arising from the superposition of a-events and b-events. Thus, in our notation, both the models are purely b-dependent processes. The marginal process generated by b-events constitutes a renewal point process. Srinivasan and Rajamannar (1970a) have provided a complete description of the marginal process by obtaining the renewal density of b-events in both the models. We shall deal with these models in detail in Chapter 9.

We next proceed to investigate the possibility of general characterization of the probabilistic structure in terms of only a partial history. As in the case of univariate point processes, a fitting characterization is provided by the sequence of generalized product densities (conditional upon a certain prehistory) of either the superposed process or the pooled process. For simplicity let us confine our attention to stationary regular processes. The first-order product densities of the bivariate process are defined by

$$(6.3.10a) \quad h_1^\alpha(t, \beta) = \lim_{\tau \to 0} \Pr\left[d_t N_\beta(0, t) \geqslant 1 \mid N_\alpha(0, \tau) = 1\right]/dt$$

or

$$(6.3.10b) \quad h_1^\alpha(t, \beta) = \lim_{\tau \to 0} E\left[d_t N_\beta(0, t) \mid N_\alpha(0, \tau) = 1\right]/dt.$$

When $\alpha \neq \beta$, $h_1^\alpha(., \beta)$ is called a "second-order cross-intensity function" (Cox and Lewis (1970)). For non-regular processes, (6.3.10b) is the appropriate definition, and in such a case it is also worthwhile to introduce a mixed density by

$$(6.3.11) \quad h_1^{\alpha\beta}(t, r)$$

$$= \lim_{\tau \to 0} \lim_{\tau' \to 0} E\left[d_t N_\beta(0, t) \mid N_\alpha(0, \tau) \geqslant 1, N_\gamma(\tau, \tau') > 1\right]/dt.$$

Regularity of a process can be characterized by the absence of delta function concentrations at the origin in the cross-intensity functions. Of course, marginal regularity will be characterized by the absence of delta function concentration at the origin in the intensity functions $h_1^\alpha(t, \alpha)$. By virtue of stationarity, we have

$$(6.3.12) \quad h_1^\alpha(t, \beta) = \sum_{n=1}^{\infty} g_n^\alpha(t, \beta),$$

where $g_n^\alpha(t, \beta)$ is to be interpreted as $f_n(t, \alpha)$ for $\alpha = \beta$. For $\alpha = \beta$, it also follows that $h_1^\alpha(t, \beta)$ is an even function of t. In the case of univariate point processes, L_{-n} and L_n are identically distributed. The corresponding result for the bivariate process is

$$(6.3.13) \quad h_1^\alpha(t, \beta)\mu_\alpha = h_1^\beta(-t, \alpha)\mu_\beta.$$

We next note that the conditioned product density of degree one of the superposed process defined by

$$(6.3.14)^{(*)} \quad h_1^{\cdot}(t, .) = \lim_{\tau \to 0} E\left[d_t N_.(0, t) \mid N_.(0, \tau) = 1\right]/dt$$

is given by

$$h_1^{\cdot}(t, .) = \sum_{\alpha, \beta} \mu_\alpha \, h_1^\alpha(t, \beta) \, (\mu_a + \mu_b).$$

The conditioned product density of degree one of the pooled as well as of the bivariate process plays an important role since the stationary second-order product density can be directly obtained from it. The analogue of (4.3.15) for the product density of degree two of events of the pooled process with

---

$^{(*)}$ We shall use the symbol "." whenever we do not wish to identify the type of the event. This convenient notation is due to Cox and Lewis (1970).

reference to an arbitrary origin is given by

(6.3.15) $h_2(x, \alpha; y, \beta) = \mu_\alpha h_1^\alpha(y - x, \beta) \quad (x < y).$

The importance of the second-order product density hardly needs to be stressed in virtue of the special role played by its Fourier transform in the evaluation or estimation of the power spectrum of the linear response to such processes (see, for example, Srinivasan and Vasudevan (1971), Chapter 5). The corresponding result for the superposed process is given by

(6.3.16) $h_2(x, .; y, .) = \sum\limits_{\alpha, \beta} \mu_\alpha h_1^\alpha(y - x, \beta) (\mu_a + \mu_b).$

It is sometimes more convenient to deal with the covariance densities rather than the second-order product densities. For instance, Bartlett (1963, 1966) always deals with covariance densities in his spectral analysis. Using the same notation as in Chapter 4, we can define

$$\gamma_{\alpha\beta}(t) = \lim_{\tau, \tau' \to 0^+} \text{cov} \ [N_\alpha(0, \tau), N_\beta(t, \tau')]/\tau\tau'$$

(6.3.17) $\qquad = \mu_\alpha h_1^\alpha(t, \beta) - \mu_\alpha \mu_\beta.$

We note that

(6.3.18) $\qquad\qquad\qquad \gamma_{\alpha\beta}(t) = \gamma_{\beta\alpha}(-t).$

For $\alpha = \beta$, $\gamma_{\alpha\beta}(.)$ is called the *autocovariance density* and can be written as

(6.3.19) $\qquad\qquad\qquad \gamma_{\alpha\alpha}(t) = \mu_\alpha \delta(t) + \gamma_{\alpha\alpha}^{\text{cont}}(t),$

where $\gamma_{\alpha\alpha}^{\text{cont}}(t)$, the continuous part of $\gamma_{\alpha\alpha}(t)$, is given by

(6.3.20) $\qquad\qquad\qquad \gamma_{\alpha\alpha}^{\text{cont}}(t) = \mu_\alpha [h_1^\alpha(t, \alpha) - \mu_\alpha].$

The covariance density of the superposed process is given by

$$\gamma_{..}(t) = \lim_{\tau, \tau' \to 0} \text{cov} \ [N_.(0, \tau), N_.(t, \tau')]/\tau\tau'$$

(6.3.21) $\qquad = \sum\limits_{\alpha\beta} [\mu_\alpha h_1^\alpha(t, \beta) - \mu_\alpha \mu_\beta],$

on the basis of which we can discuss the correlational structure of the bivariate process.

The *covariance time surface* which represents the covariance between $N_\alpha(0, t_1)$ and $N_\beta(0, t_2)$ is given by

$$V^{\alpha\beta}(t_1, t_2) = \text{cov} \left[\int\limits_0^{t_1} d_u N_\alpha(0, u), \int\limits_0^{t_2} d_v N_\beta(0, v) \right]$$

(6.3.22) $\qquad = \int\limits_0^{t_1} \int\limits_0^{t_2} \gamma_{\alpha\beta}(u - v) \ du \ dv.$

If we specialize to the case where $t_1 = t_2 = t$, we obtain the variance function a

$$(6.3.23) \quad V^{\alpha\beta}(t) \equiv V^{\alpha\beta}(t,t) = \int_0^t (t-v)[\gamma_{\alpha\beta}(v) + \gamma_{\beta\alpha}(v)]\,dv,$$

which provides a generalization of equation (4.3.18). By definition it follows that

$$(6.3.24) \qquad\qquad\qquad V^{\alpha\alpha}(.) > 0.$$

It can also be shown (see problem 6.3, below) that

$$(6.3.25) \quad [V^{ab}(t_1, t_2)]^2 \leqslant V^{aa}(t_1)\, V^{bb}(t_2),$$

so that any set of three functions cannot form covariance densities. It would be interesting to demonstrate the existence of a bivariate point process for any set of three functions $V^{ab}(.,.)$, $V^{aa}(.)$, $V^{bb}(.)$, satisfying (6.3.24) and (6.3.25).

6.4 Interacting renewal processes

For purposes of illustration, let us consider a pair of interacting renewal processes. The motivation for the study of such a bivariate process is due to a model proposed by ten Hoopen and Reuver (1965) in their study of neurophysiological processes. The model was subsequently studied in detail by Srinivasan and Rajamannar (1970b, c) and Lawrance (1970), who dealt with the properties of the marginal processes. One of the pair of processes is assumed to be a stationary recurrent process (known as the "inhibitory process"), hereinafter called the series of a-events, the interval between any two successive a-events being governed by the probability density function $\phi(.)$. The other member of the pair is the point process of b-events which are obtained by a procedure of selection, or filtered from another independent stationary recurrent process (known as the "excitatory process") in which the intervals between successive events are governed by the common probability density function $\psi(.)$. The selection or filtration of events is achieved by allowing every inhibitory to delete the next excitatory event. If one or more inhibitories occur before the next excitatory event following an inhibitory event, only that particular excitatory event is deleted, so that the next excitatory following the deleted one is undeleted. Fig. 6.2 shows a set of typical sequences

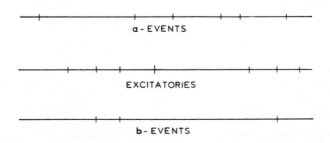

Fig. 6.2   The events of a special bivariate process plotted on the time axis

of inhibitories, excitatories and undeleted excitatories. Ten Hoopen and Reuver (1965, 1967b) studied the process by dealing with the probability density function governing the interval between any two successive undeleted events, while Srinivasan and Rajamannar dealt with the conditioned product density of degree one and the power spectrum of undeleted events. Lawrance (1970), by a direct and elaborate analysis of the properties of the distribution function of successive intervals, obtained an explicit expression for the expected number of undeleted events in an arbitrary interval $(0, t)$. We shall now attempt to characterize the process by the generalized renewal densities, or equivalently by conditioned product densities.

We shall consider a more general situation by allowing the excitatories to be a stationary point process which is not necessarily regular. This would mean that if a bunch of excitatories occur which are susceptible to inhibition, only one of them is deleted, the rest contributing towards the b-events which form a marginally irregular point process. The complete intensity functions $\lambda_\alpha(t \mid \mathcal{H}_t)$ depend on the inhibitories only to the extent of the backward recurrence time of the inhibitory,

The special bivariate point process can be characterized by the semi-synchronous renewal densities of the pooled process as defined by (6.3.10b). The densities can be directly obtained by making use of the mechanism of selective interaction. First let us consider the case when $\beta = b$. As noted earlier, the generalized factorial moment density of order 2 is easier to deal with. Defining $h_2^f(t_1, \alpha; t_2, \beta)$ as the factorial moment density corresponding to the product density $h_2(t_1, \alpha; t_2, \beta)$, we note in view of stationarity that

$$(6.4.1) \quad h_2^f(t_1, \alpha; t_2, \beta) = \mathcal{F}^\alpha(t_2 - t_1; \beta) \quad t_2 > t_1,$$

where $\mathcal{F}^\alpha(x, \beta)$ is interpreted as the factorial moment density corresponding to an event of type α at the origin and another event of type β at x $(x > 0)$. To obtain $\mathcal{F}(x, b)$ for $x > 0$, we note that synchronization of a b-event with the origin implies that no a-event has occurred in the immediate past beyond the origin until the previous e-event, which may or may not be a b-event. Thus we need the history of the process prior to the origin only up to $-u$ $(u > 0)$, the point synchronizing with the occurrence of an e-event. It is also clear that the point $-u$ is an arbitrary point so far as the b-process is concerned. We classify the collection of events[*] leading to a b-event in the interval $(x, x + dx)$ into two mutually exclusive classes according as whether or not an a-event occurs in $(0, x)$. In the former case, non-zero contributions to the event in question can arise only from those configurations in which no a-event occurs in $(-u, x)$, u being an arbitrary point. If, on the other

---

[*] The word "event" is used in a general probabilistic sense and does not refer to the events of the point process.

hand, an a-event occurs in $(0, x)$, the last a-event in $(0, x)$ should be followed by at least one a-event, so that the e-event that occurs in $(x, x + dx)$ is a b-event. Adding these contributions, we obtain

$$(6.4.2) \quad \mathcal{F}_1^b(x, b) = \mu_e \int_0^\infty du \int_{x+u}^\infty g_1(v, a) fr_{1,1}(u, x + u; e) dv$$

$$+ \mu_e \int_0^\infty du \int_u^{u+x} dv \, g_1(v, a) \int_{x+u}^\infty \phi(\bar{v} - v) d\bar{v} \int_{v-u}^x fr_{2,1}(u, u + w, u + x; e) dw$$

$$+ \mu_e \int_0^\infty du \int_u^{u+x} dv \, g_1(v, a) \int_v^{u+x} h^a(\bar{v} - v) d\bar{v} \int_{x+u}^\infty \phi(w - \bar{v}) dw$$

$$\times \int_{\bar{v}-u}^x fr_{2,1}(u, u + \bar{w}, u + x; e) d\bar{w},$$

where $g_1(., a)$ is the probability density function governing the forward recurrence time of a-events and $h^a(.)$ is the ordinary renewal density of a-events. The functions $fr_{..}(..., e)$ are mixed factorial moment densities of e-events defined by

$$(6.4.3) \quad fr_{1,1}(x, y; e) = \lim_{\Delta_1, \Delta_2, \Delta_3 \to 0} \Pr \{ N_e(0, x) = 0, N_e(x, \Delta_1) = 1,$$

$$N_e(y, \Delta_2) = 1 \mid N_e(-\Delta_3, \Delta_3) = 1 \} / \Delta_1 \Delta_2;$$

$$(6.4.4) \quad fr_{2,1}(x, y, z; e)$$

$$= \lim_{\substack{\Delta_1, \Delta_2 \to 0 \\ \Delta_3, \Delta_4 \to 0}} \Pr [N_e(0, x) = 0, N_e(x, \Delta_1) = 1, N_e(y, \Delta_2) = 1,$$

$$N_e(y + \Delta_2, z - y - \Delta_2) = 0,$$

$$N_e(z, \Delta_3) = 1 \mid N_e(-\Delta_4, \Delta_4) = 1] / \Delta_1 \Delta_2 \Delta_3.$$

In the case of e-events forming a regular renewal process, the mixed factorial moment densities are given by

$$(6.4.5) \quad fr_{1,1}(x, y; e) = f^e(x) h^e(y - x) \quad (x < y)$$

$$(6.4.6) \quad fr_{2,1}(x, y, z; e) = f^e(x) h^e(y - x) f^e(z - y) \quad (x < y < z),$$

where $f^e(.)$ is the probability density function governing the interval between any two successive e-events and $h^e(.)$ is the ordinary renewal density of e-events.

The stationary value of the intensity of b-events can be directly calculated by tracing the history backwards from an arbitrary b-event to the previous e-event. Thus we have

$$\mu_b = \mu_e \int_0^\infty du \int_u^\infty g_1(v, a) f^e(u) dv$$

$$(6.4.7) \quad = \mu_e \mu_a \int_0^\infty f^e(u) du \int_u^\infty dv \int_v^\infty \phi(w) dw,$$

or equivalently

(6.4.8) $\mu_b = \mu_e - \mu_a + \mu_a \int_0^\infty \phi(y) \Pr[N_e(0, y) = 0] \, dy,$

a result obtained by Lawrance (1970). Thus the average number of b-events in an arbitrary interval $(0, t)$ is given by

(6.4.9) $E[N_b(0, t)] = \mu_b t.$

It is interesting to note that (6.4.9) taken with (6.4.7) is true even if the inhibitory events (a-events) form a stationary point process (which need not again be regular). Next we note that the generalized synchronous renewal density is given by

(6.4.10) $r_1^b(x, b) = \mathcal{F}_1^b(x, b)/\mu_b.$

So far we have assumed that the a-events form a renewal point process. To extend formula (6.4.3) corresponding to the stationary stream of a-events, we introduce the random variable L_n representing the length measured from an arbitrary point to the nth a-event. Formula (6.4.3) is valid in the general case if we replace $g_1(v, a) \phi(\bar{v} - v)$ by $\mathcal{G}(v, \bar{v}, a)$ and $g_1(v, a) h^a(\bar{v} - v) \phi(w - \bar{v})$ $\sum_{n=2}^\infty \mathcal{G}_{1,n,n+1}(v, \bar{v}, w, a)$, where

(6.4.11) $\mathcal{G}_{1,2}(v, \bar{v}, a) = \Pr[v < L_1 < v + dv, \bar{v} < L_2 < \bar{v} + d\bar{v}]/dv \, d\bar{v},$

(6.4.12) $\mathcal{G}_{1,n,n+1}(v, \bar{v}, w, a)$
$= \Pr[v < L_1 < v + dv, \bar{v} < L_n < \bar{v} + d\bar{v}, w < L_{n+1} < w + dw]/dv \, d\bar{v} \, d$

The pure covariance density of b-events as defined by Bartlett (1966) is given by

(6.4.13) $\gamma_b^b(x) = \lim_{\substack{\Delta \to 0 \\ \Delta' \to 0}} E\left[\dfrac{N_b(t, t + \Delta) N_b(t + x, t + x + \Delta')}{\Delta \Delta'} - \mu_b^2\right] \quad x > 0,$

while the complete covariance density is given by

(6.4.14) $\gamma_b^b(x)^{\text{complete}} = \gamma_b^b(x) + \delta(x) \sum k^2 \lambda_k \ (b\text{-event})$
$= \mu_b \delta(x) + \mathcal{F}_1^b(x, b) - \mu_b^2.$

It should be noted that $\mathcal{F}_1^b(x, b)$ is not continuous and contains a delta function concentration at the origin. Of course, when the process is regular, $\mathcal{F}_1^b(x, b)$ is continuous. Thus a complete spectral analysis of b-events is possible with the help of equation (6.4.2).

Next we note that the mean square number of b-events can be related to $\mathcal{F}_1^b(., b)$ by the following relation (which follows by definition):

(6.4.15) $E[N_b(0, t)]^2 = E[N_b(0, t)] + 2\int_0^t (t - u) \mathcal{F}_1^b(u, b) \, du.$

We next proceed to other types of densities implied by (6.3.10b). The generalized renewal density $r_1^a(x, a)$ is simply the ordinary renewal density of a-events since it is independent of b-events. The semi-synchronous cross-densities $r_1^\alpha(x, \beta)$ $(\alpha \neq \beta)$ can be obtained without recourse to the corresponding factorial moment density.

To obtain $r_1^b(x, a)$ we note that the collection of events can be broadly classified into two categories according as whether an a-event occurs in the interval $(0, x)$ or not. In either case we note that the origin is an arbitrary point for the marginal process of e-events. If no a-event occurs in $(0, x)$, then a non-zero contribution to the density arises from the class of events which give rise to at least one e-event in $(0, x)$, apart from the one in $(x, x + dx)$. If, on the other hand, an a-event occurs, a non-zero contribution to the density arises again from the class of events which give rise to at least one e-event subsequent to the a-event. Adding these two contributions, we find that

$$(6.4.16) \quad r_1^b(x, a)$$

$$= \mu_e \int_0^x r_1(x - u, e)\, du \int_x^\infty \phi(v)\, dv + \mu_e \int_0^x h^a(u)\, du \int_{x-u}^\infty \phi(\bar{v})\, d\bar{v} \int_u^x r_1(x - w, e)\, dw$$

or

$$(6.4.17) \quad r_1^b(x, a)$$

$$= \mu_e H(x, e) \int_x^\infty \phi(v)\, dv + \mu_e \int_0^x h^a(u)\, H(x - u, e)\, du \int_{x-u}^\infty \phi(\bar{v})\, d\bar{v},$$

where $H(., e)$, $r(., e)$ and $h^a(.)$ are respectively the generalized renewal function of e-events, the generalized renewal density of e-events, and the ordinary renewal density of a-events.

To find an explicit expression for $r_1^a(x, b)$, we note that we trace the past history of the process up to the e-event prior to the b-event at the origin and argue as in the earlier case. Thus, using Bayes' theorem, we find that

$$(6.4.18) \quad r_1^a(x, b) = \int_0^\infty f^e(u)\, g_1(u + x, a)\, du \Big/ \int_0^\infty f^e(u)\, du \int_u^\infty g_1(v, a)\, dv,$$

where $f^e(.)$ is the p.d.f. governing the length between an arbitrary e-event and the following e-event. Again we note that (6.4.18) is valid even when the a-events form a stationary point process.

Next we note that the generalized renewal density of the superposed process of a- and b-events is given by

$$(6.4.19) \qquad r_1(x, .) = \sum_{\alpha, \beta} r_1^\alpha(x, \beta),$$

with the help of which a spectral analysis can be carried through based on the spectral properties of a- and e-events.

Finally we shall consider two special cases for purposes of illustration.

At the outset we note that if the e-events form a regular renewal process, the marginal process of b-events is a regular renewal process, provided the a-events form a simple Poisson process. In such a case we note that (6.4.2) simplifies to

$$(6.4.20) \quad \mathcal{F}_1^b(x, b)$$

$$= \mu_e \int_0^\infty f^e(u) \, e^{-\lambda_a u} \, du \, h^e(x) \, e^{-\lambda_a x} + \mu_e \, e^{-\lambda_a x} \int_0^\infty f^e(u) \, e^{-\lambda_a u} \, du$$

$$\times \left[\lambda_a \int_0^x dv \int_v^x h^a(w) \, \phi(x - w) \, dw + \lambda_a^2 \int_0^x e^{-\lambda_a v} \, dv \int_v^x e^{\lambda_a \bar{v}} \, d\bar{v} \int_v^x h^a(w) \, \phi(x - w) \, dw \right]$$

where λ_a is the parameter characterizing the Poisson a-events. If $r_1^{b*}(s, b)$ is the Laplace transform of $r_1^b(x, b)$, using (6.4.10), (6.4.8) and (6.4.20), we find that

$$(6.4.21) \quad r_1^{b*}(s, b) = [h^{e*}(s) - h^{e*}(\lambda_a + s)] f^{e*}(s)$$

or

$$(6.4.22) \quad r_1^{b*}(s, b) = f^{e*}(s) \left[\frac{f^{e*}(s)}{1 - f^{e*}(s)} - \frac{f^{e*}(s + \lambda_a)}{1 - f^{e*}(s + \lambda_a)} \right].$$

Since the b-process is a renewal process, it is completely characterized by the ordinary renewal density whose Laplace transform is given by (6.4.22). From this equation we can also obtain the Laplace transform of the probability density governing the interval between an arbitrary b-event and the next b-event. Using the results of renewal theory, we obtain from (6.4.22)

$$(6.4.23) \quad f^{b*}(s) = \frac{f^{e*}(\lambda_a + s)}{1 - f^{e*}(s) + f^{e*}(\lambda_a + s)},$$

a result first proved by ten Hoopen and Reuver. Although the b-events form a renewal process, the intervals of the bivariate process consisting of a- and b-events are not independently distributed. Thus the bivariate process cannot be uniquely characterized by the generalized renewal density $r_i(x, .)$. On the other hand, we need the sequence of product densities of the type $h_n(x_1, ., x_2, ., \ldots, x_n, .)$ $(n = 1, 2, \ldots)$ to characterize the process completely. These densities can be obtained through the corresponding sequence of densities of the pooled process $h_n(x_1, \alpha_1; x_2, \alpha_2; \ldots; x_n, \alpha_n)$. If $\alpha_1 = \alpha_2 = \ldots = \alpha_n = a$, we simply have

$$(6.4.24) \quad h_n(x_1, \alpha_1; x_2, \alpha_2; \ldots; x_n, \alpha_n) = \lambda_a^n.$$

On the other hand, if $\alpha_i = b$ $(i = 1, 2, \ldots, n)$, by the renewal nature of the marginal b-events we have for $x_1 < x_2 < \ldots < x_n$

$$(6.4.25) \quad h_n(x_1, b; x_2, b; \ldots; x_n, b)$$

$$= \mu_b \, r_1^b(x_2 - x_1, b) \, r_1^b(x_3 - x_2, b) \ldots r_1^b(x_n - x_{n-1}, b),$$

where $r_1^b(.,b)$ is defined by (6.4.11) with $\mathcal{F}_1^b(.,b)$ given by (6.4.20). If, on the other hand, we have a general mixture of a- and b-events, let us assume that there are k b-events $(k < n)$ occurring at $x_{i_1}, x_{i_2}, \ldots, x_{i_k}, i_1, i_2, \ldots, i_k$ being positive integers such that $0 < i_1 < i_2 < \ldots < i_k < n$. The Poisson nature of the a-events and the renewal nature of the e-events imply that $x_{i_1}, x_{i_2}, \ldots, x_{i_k}$ are the *points of regeneration* (see, for example, Palm (1943) and Bellman and Harris (1948)). Then it is clear that for $x_1 < x_2 < \ldots < x_n$, we have

(6.4.26) $\quad h_n(x_1, \alpha_1; x_2, \alpha_2; \ldots ; x_n, \alpha_n)$

$$= \lambda_a^{i_1 - 1} r_1^b(x_{i_1} - x_{i_1-1}, a)(\lambda_a)^{n-i_k} \prod_{j=1}^{k-1} \mathcal{R}(x_{i_{j+1}} - x_{i_j}, x_{i_j+1} - x_{i_j}),$$

where $\alpha_{i_j} - b$ $(j = 1, 2, \ldots, k)$ and $\alpha_m = a (m \notin [i_1, i_2, \ldots, i_k])$ and $r_1^b(.,a)$ is given by (6.4.17). The function $\mathcal{R}(.,.)$ is defined by

(6.4.27) $\quad \mathcal{R}(x_{i_{j+1}} - x_{i_j}, x_{i_j+1} - x_{i_j})$

$$= \Pr\,[dN_b(e, x_{i_{j+1}}) = 1,$$
$$dN_a(0, x_m) = 1 \;(i_j < m < i_{j+1}) \,|\, dN_b(0, x_{i_j}) = 1]/ \prod_{l=i_j+1}^{i_{j+1}} dx_l.$$

Using the regenerative nature of the b-events and arguments similar to those in the derivation of (6.4.17) or (6.4.18), we obtain for $i_{j+1} > i_j + 1$

(6.4.28) $\quad \mathcal{R}(x_{i_{j+1}} - x_{i_j}, x_{i_j+1} - x_{i_j})$

$$= \lambda_a^{i_{j+1} - i_j} \int_{x_{i_j+1}}^{x_{i_{j+1}}} h^e(u - x_{i_j}) f^e(x_{i_{j+1}} - u) \exp\{-\lambda_a(x_{i_{j+1}} - u)\}\,du,$$

while for $i_{j+1} = i_j + 1$ we have the relation

(6.4.29) $\quad \mathcal{R}(x_{i_{j+1}} - x_{i_j}) = f^e(x_{i_{j+1}} - x_{i_j}) \exp\{-\lambda_a(x_{i_{j+1}} - x_{i_j})\}$

$$+ \int_{x_{i_j}}^{x_{i_{j+1}}} h^e(u - x_{i_j}) f^o(x_{i_{j+1}} - u) \exp\{-\lambda_a(x_{i_{j+1}} - u)\}\,du.$$

The product density (or generalized renewal density) of the bivariate process of degree n is given by

(6.4.30) $\quad h_n(x_1., x_2., \ldots, x_n.) = \sum_{\alpha_1, \alpha_2, \ldots, \alpha_n} h_n(x_1, \alpha_1; x_2, \alpha_2; \ldots ; x_n, \alpha_n).$

Thus a complete characterization of the bivariate process is provided by (6.4.24) and (6.4.25)–(6.4.30), with the help of which the complete spectral properties of the bivariate process can be studied.

We next consider the case when the a-events form a stationary renewal process and the e-events a simple Poisson process with parameter λ_e. In this case the b-events form a non-renewal point process. Therefore the probability density of the interval between any two b-events is not sufficient to characterize the marginal distribution of b-events. Recently Lawrance (1971) has presented an analysis of the distribution of several consecutive intervals of b-events in order to arrive at the second-order spectral properties of the marginal process. At the outset, we note that the stationary product density of the b-events is given by (6.4.3), provided we make the appropriate substitutions. In the special case under consideration, (6.4.2) becomes

$$(6.4.31) \quad \mathcal{F}(x, b) = \lambda_e^2 \int\limits_0^\infty e^{-\lambda_e u}\, du \int\limits_{x+u}^\infty g_1(v, a) \lambda_e \, dv$$

$$+ \lambda_e^2 \int\limits_0^\infty e^{-\lambda_e u}\, du \int\limits_u^{u+x} g_1(v, a)\, dv \int\limits_{x+u}^\infty \phi(\bar{v} - v)\, d\bar{v} \int\limits_{v-u}^x \lambda_e^2\, e^{-\lambda_e(x-w)}\, dw$$

$$+ \lambda_e^2 \int\limits_0^\infty e^{-\lambda_e u}\, du \int\limits_u^{u+x} g_1(v, a)\, dv \int\limits_v^{u+x} h^a(\bar{v} - v)\, d\bar{v} \int\limits_{x+u}^x \phi(w - \bar{v})\, dw$$

$$\times \int\limits_{\bar{v}-u}^x e^{-\lambda_e(x-\bar{w})}\, d\bar{w}.$$

Using (6.4.9), we can then identify μ_b:

$$(6.4.32) \quad \mu_b = \lambda_e - \lambda_a + \lambda_a \int\limits_0^\infty \phi(u)\, e^{-\lambda_e u}\, du.$$

Thus we can deal with the generalized synchronous renewal density of b-events or, better, its Laplace transform. Hence we have, after some straightforward calculation,

$$(6.4.33) \quad r_1^{b*}(s, b) = \int\limits_0^\infty r_1^b(x, b)\, e^{-sx}\, dx$$

$$= \frac{\lambda_e}{s} - \frac{\lambda_e^2}{\lambda_e + s} \frac{[1 - \phi^*(\lambda_e + s)]}{(1 - \phi^*(s))[s - \lambda_e]} \frac{[g_1^*(\lambda_e, a) - g_1^*(s, a)]}{(1 - g_1^*(\lambda_e, a))}$$

with

$$(6.4.34) \qquad\qquad g_1^*(s, a) = \frac{\lambda_a}{s}[1 - \phi^*(s)].$$

The above result was obtained by Srinivasan and Rajamannar (1970a) as a special case of the general result when both a- and e-events form independent renewal processes. The extension to the nth order product density appears to be difficult, although product densities of the first few orders of b-events can be written down using arguments similar to those in the derivation of (6.4.2). The special density of the pooled point process can be obtained using (6.4.19), since explicit expressions for the renewal densities of the bivariate process are available through (6.4.17) and (6.4.18).

6.5 Examples

In this section we shall cite three examples from the area of Management Science. The reader who is not familiar with the subject is advised to go through Chapter 8 of this monograph. The use of multivariate analysis has not been examined in any depth in the literature so far. The approach through multivariate analysis is full of promise, and we hope the following paragraphs will demonstrate the usefulness of such an approach in Management Science.

6.5.1   Theory of queues (single server system)

A simple example of a bivariate point process is provided by the point process consisting of the events corresponding to the arrivals and departures in a queuing process. A queuing system is completely described by (a) the input process, (b) the queue discipline, and (c) the service mechanism. The input describes the manner in which the customers arrive and join the queue. Suppose customers arrive at the counter at times $t_1, t_2, \ldots$, and let $X_n = t_{n+1} - t_n$ denotes the time-interval between the arrivals of the nth and $(n + 1)$th customers. The input process is given by the probability law governing the sequence of arrival times t_i or the sequence of the inter-arrival times X_n. Queue discipline is the rule determining the formation of the queue, the simplest rule being "first come, first served", i.e. customers are served in order of arrival. The service mechanism spells out the probability law governing the sequence of service times τ_n corresponding to the nth customer. In what follows we shall assume that there is a single counter offering service and that the service times of the customers are all independently and identically distributed with p.d.f. $\phi(.)$. Likewise the inter-arrival times X_n will be assumed to be independently and identically distributed with p.d.f. $f(.)$. Let the sequence of a-events correspond to the sequence of the epochs of arrivals and d-events to the epochs of departures.

By the assumption of independence of inter-arrival times, the a-events form a renewal process. Thus there is no need to discuss the marginal a-events. On the other hand, the marginal process of d-events constitutes a general point process. The first-order conditioned product densities of the bivariate process will describe the correlational structure and the spectral properties. It may be useful to characterize the process (to second order) by means of functions of the type $h_1^\alpha(t, \beta)$ $(\alpha, \beta = a, d)$ as defined by (6.3.10a) or (6.3.10b). Sometimes it is quite useful to characterize the type of a-events that may give rise to b-events of certain specified properties (see problem 6.5, below).

6.5.2   Theory of dams with infinite capacity

The theory of dams is the description of a special type of inventory problem in which the level of the inventory (namely the height of water-level of the dam) is a continuous variable. There are several models of dam, depending

118

on the nature of the input and the release policy. For purposes of illustration
let us assume the release rate to be a constant that is equal to unity as long
as the water-level $W(t)$ at any time t is greater than zero, there being no release when $W(t) = 0$. The input is assumed to occur at different epochs, the
time-intervals between successive inputs being independently and identically
distributed with p.d.f. $f(.)$. Likewise the quantities of input at different epoc
are assumed to be independently and identically distributed with p.d.f. $\phi(.)$.
These assumptions provide a model of a dam with no restriction on its capacity. An interesting bivariate point process arises if we consider the zero

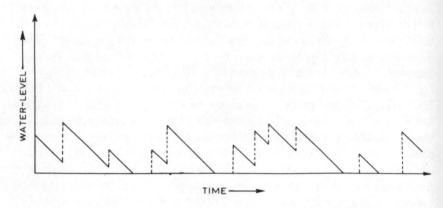

Fig. 6.3   The water-level of a dam plotted against time

crossings of $W(t)$. A typical plot of $W(t)$ is shown in Fig. 6.3. The epochs at
which $W(t)$ drops to zero from above are the epochs of commencement of
"dry spells". We shall call such events a-events, and the events corresponding
to the epochs of the termination of the dry spell b-events. It is clear from
the assumption that the marginal process of b-events is a renewal process,
provided that $\int_0^\infty tf(t)\,dt/\int_0^\infty t\phi(t)\,dt > 1$. (For a proof see Prabhu (1965).) The
intensity functions of the stationary bivariate process of a- and b-events can
be obtained from the function $\pi(x, y, t)$, where

$$\pi(x, y, t) = \lim_{\substack{\Delta \to 0 \\ \Delta' \to 0}} \Pr\{x < W(t) < x + \Delta \mid y = W(0) > W(-\Delta')\}/\Delta$$

(see problem 6.6, below). If an upper limit k is placed on the capacity of
the dam, then overflows are possible. In such a case it is possible to define
events of type c corresponding to epochs of overflow. It is interesting to
study the properties of the multivariate point process of a-, b- and c-events.

6.5.3 A reliability model

Let us consider a system consisting of two identical units (machines) subject to failure. It is assumed that the failure-free periods of the units are independently and identically distributed with p.d.f. $f(.)$. Initially, one unit is put into operation and the other is kept on cold-standby. A unit on cold-standby does not fail with probability one. As soon as the unit that is put into operation fails, the unit on cold-standby is put into operation and the unit that has failed is taken up for repair by a repair facility. The repair time of such a unit is assumed to be governed by the p.d.f. $\phi(.)$. If the repair facility completes the repair of the unit while the system is still functioning, the unit is kept as standby. If, however, the unit that is functioning fails before the repair facility completes the repair of the unit that has already failed, then the unit waits as if in a queue until the repair facility becomes free. In such a case the system ceases to function until the unit undergoing repair becomes available for operation. The service facility takes up the other unit for repair as soon as it becomes free. In general, it is possible for the system to pass through cycles of operative and inoperative periods known as the "up" and "down" periods. The state of the system can be described by a discrete stochastic process $X(t)$ with the state space (see Chapter 1) consisting of three points 0, 1 and 2 specified as follows:-

Value of the state variable	State of the system
0	One unit in operation and one unit on standby.
1	One unit in operation and one unit under repair.
2	One unit undergoing repair and the other waiting in queue at the repair facility.

Let us consider the point process generated by the epochs of entry into the states 0, 1 and 2. If we denote the events by a, b, c respectively, corresponding to states 0, 1 and 2, then the system can be studied from the viewpoint of the multivariate point process. By shifting the time-origin to minus infinity, we can generate a stationary multivariate point process whose characteristics can be described by the multivariate product densities of a, b and c events. Srinivasan and Gopalan (1974) have obtained the entrance probabilities of the states 0, 1 and 2, and by a limiting procedure they obtain the stationary second-order bivariate product densities of a, b and c events. It is possible to extend these ideas to an n-unit system.

6.1   Consider three independent regular point processes, say a, b and c. Events of type I are obtained by superposing a- and b-events, while events of type II are obtained by superposing b- and c-events. Show that the bivariate process of events of types I and II is marginally regular but not regul[ar]

6.2   Show that if $T(1)$ represents the type of the first event counted from the origin, then

$$\Pr\{x < L_1(a) < x + dx, \ T(1) = a, \ L_1(b) - L_1(a) > y\}/dx$$

$$= \int_{y+x}^{\infty} g_1^{ab}(x, u) \, du$$

$$= \mu_a \Pr\{L_{-1}^a(.) > x, L_1^a(b) > y\}.$$

Deduce that

$$\mu_a \, \mathrm{E}\{[L_1^a(b)]^r\} + \mu_b \, \mathrm{E}\{[L_1^b(a)]^r\}$$

$$= \mu_a \, \mathrm{E}\{|L_1^a(b) - L_1^a(a)|^r\} + \mu_b \, \mathrm{E}\{|L_1^b(b) - L_1^b(a)|^r\}.$$

(Wisniewski, 1972)

6.3   Show that, for all values of p and q,

$$\mathrm{var}\{pN_a(0, t_1) + qN_b(0, t_2)\} \geqslant 0.$$

Deduce that

$$[V^{ab}(t_1, t_2)]^2 < V^{aa}(t_1) \, V^{bb}(t_2).$$

6.4   Consider an arbitrary regular stationary process of events of type a. Let each event of type a be subject to a random displacement to form a corresponding event of type b, the displacements of different points being independently and identically distributed with probability density function $p(.)$; denote the probability density function of the difference between two such random variables as $q(.)$. Obtain the integral equations satisfied by the intensity functions $h_1^b(t, a)$ and $h_1^b(t, b)$. Deduce that if the type a-events form a Poisson process, then

$$h_1^b(t, a) = p(t) + \lambda_a$$

$$h_b^b(t) = \lambda_a,$$

where λ_a is the intensity of the a-events.

6.5   Show that the d-events of the queuing process (as defined in section *6.5.1*) form a Poisson process if and only if the a-events form a Poisson process and the service times are exponentially distributed.   (Daley, 1968)

6.6   Show that the stationary intensity functions $h_1^b(., \alpha)$ $(\alpha = a, b)$ of the bivariate process defined in section 6.5.2 are given by

$$h_1^b(t, a) = \lim_{x \to 0} \int_0^\infty \pi(x, y, t) \, \phi(y) \, dy$$

$$h_1^b(t, b) = \lim_{x \to 0} \int_0^t du \, g_1(u) \int_0^\infty \pi(x, y, t - u) \, \phi(y) \, dy.$$

6.7   Show that the marginal process of c-events defined in section 6.5.2 is a renewal process. Show also that the multivariate point process of a, b and c events is irregular, although it is marginally regular.

7   STATISTICAL PHYSICS

7.1   Introduction

The statistical theory of equilibrium states of matter as developed by
Boltzmann, Jeans, Lorentz and Gibbs deals with a large assembly of similar
molecules; their material or thermodynamic properties being interpreted in
terms of the averages of certain appropriate quantities. The stability of the
material properties is a consequence of the fundamental postulate of Statistical Physics which ensures that the mean values (corresponding to the different material properties) are the true values by virtue of the negligible natur
of the fluctuations. This fundamental postulate of statistical physics has never
been in doubt inasmuch as it is not in conflict with the Gibbs *ansatz* for the
energy distribution. However, it may be worth while to investigate whether
the postulate can be deduced as a consequence of the properties of a statistical assembly of molecules. In fact, this problem attracted the attention of
Khintchine (1960) who identified the phenomenological quantities defined
in thermodynamics as the *sum functions* that possess the normality property.
The general theorems of statistical physics proved by Khintchine do provide
an air of finality for the statistical theory of equilibrium states of matter.
One should, however, consider the conditions under which the fluctuation
theorem of statistical physics holds. The physical properties of matter not
in equilibrium cannot be explained on the basis of the limit theorems of
Khintchine; for instance, the problems of phase transition and of the macroscopic motion of gases and liquids cannot be covered by limit theorems. It
is precisely in such realms that the application of the theory of point processes may help to elucidate some of the fundamental problems. Many
later contributions have been made in this direction, particularly by Cohen
(1968), Ichimaru (1968, 1970), Srinivasan (1966, 1967, 1971) and Uhlenbeck
(1963, 1968). In fact, some of the early attempts at a heuristic treatment of
point processes are due to kinetic theorists (see Yvon (1937), Bogoliubov
(1962)) who dealt with the distribution of particles over a continuous parameter. Product densities and other generic distribution functions were introduced in the process of elucidation of the fluctuation phenomena associated
with such processes. In this chapter we shall deal with some of the problems
of statistical physics that can be described in terms of point processes. The
layout of the chapter is as follows. We first deal with the formulation in
Statistical Mechanics of a system of classical particles by considering point

processes in phase space. In section **7.3** we examine the non-equilibrium properties of matter. The derivation of hydrodynamic equations and its implications are discussed, with special reference to the general fluctuation theorem and transition phenomena. We then outline a semi-classical approach to Quantum Statistics and provide a derivation of the formulae relating to the energy spectra of particles obeying different statistics.

7.2 Statistical physics of matter in equilibrium

Let $N(\mathbf{p}, \mathbf{q}, t)$ be the number of molecules that are found at time t contained in the spherical part of the phase space with origin at centre and $|(\mathbf{p}, \mathbf{q})|$ as radius, so that $dN(\mathbf{p}, \mathbf{q}, t)$ represents the number contained in the spherical shell corresponding to the momentum and coordinate lying respectively between $\mathbf{p}$ and $\mathbf{p} + d\mathbf{p}$ and $\mathbf{q}$ and $\mathbf{q} + d\mathbf{q}$. We shall adopt the Gibbsian *grand ensemble* approach in assuming that the number of particles as well as the total energy are random variables. It should be noted that beyond this we do not make any assumptions[*], and that our object is to deduce the distribution laws and provide a proof of Gibbs' theorem concerning the stability of the mean values of various macroscopic quantities, including the total energy. Thus $dN(\mathbf{p}, \mathbf{q}, t)$ is a random variable, and, following the standard procedure of stochastic point processes, let us assume that

$$(7.2.1) \quad \begin{cases} \Pr\{dN(\mathbf{p}, \mathbf{q}, t) = 1\} = f_1(\mathbf{p}, \mathbf{q}, t)d\Omega + o(d\Omega) \\ \Pr\{dN(\mathbf{p}, \mathbf{q}, t) = n\} = o(d\Omega) \quad n > 1, \end{cases}$$

so that we have

$$(7.2.2) \quad E\{dN(\mathbf{p}, \mathbf{q}, t)\} = f_1(\mathbf{p}, \mathbf{q}, t)d\Omega,$$

where $d\Omega$ is an infinitesimal element of the phase space of a single molecule. The function $f_1(\mathbf{p}, \mathbf{q}, t)$ is the product density of degree one defined over the phase space Ω and can also be interpreted as the differential mean number indicated by equation (7.2.2). Higher-order product densities arise if we evaluate the expectation value of the product of random variables of the type $dN(\mathbf{p}_1, \mathbf{q}_1, t) \, dN(\mathbf{p}_2, \mathbf{q}_2, t) \ldots dN(\mathbf{p}_n, \mathbf{q}_n, t)$.

We next observe that assumption (7.2.1) is reasonable for particles obeying Fermi statistics. If, on the other hand, the particles obey Bose or Boltzmann statistics, then the probability that $dN(\mathbf{p}, \mathbf{q}, t) > 1$ is of the same order of magnitude as the probability that $dN(\mathbf{p}, \mathbf{q}, t) = 1$. In such a case, we have to use the multiple product-density technique (see Chapter 2). However, for the purpose of our present discussion, it is not necessary to introduce explicitly the techniques of multiple point processes since $f_1(\mathbf{p}, \mathbf{q}, t)$ can still

[*] In the Gibbsian approach an additional assumption is usually made regarding the probability distribution of the total energy of the system.

124

be interpreted as the differential mean number in Ω-space, even on the understanding that multiple points have non-zero probabilities.[*]

The basic quantities like mass (number), momentum, pressure tensor, and local energy (temperature) of a small drop of fluid are respectively given by

(7.2.3) $M = \int m \, dN(\mathbf{p}, \mathbf{q}, t)$

(7.2.4) $\mathbf{P} = \int \mathbf{p} \, dN(\mathbf{p}, \mathbf{q}, t)$

(7.2.5) $\mathscr{P}_{ij} = \int m\left(\dfrac{p_i}{m} - u_i\right)\left(\dfrac{p_j}{m} - u_j\right) dN(\mathbf{p}, \mathbf{q}, t)$

$\qquad\qquad + \tfrac{1}{2} \int\int F_1^{12}(\mathbf{q}_{12}) \, q_{12j} \, dN(\mathbf{p}_1, \mathbf{q}_1, t) \, dN(\mathbf{p}_2, \mathbf{q}_2, t)$

(7.2.6) $Q = \int \tfrac{1}{2} m \left(\dfrac{\mathbf{p}}{m} - \mathbf{u}\right)^2 dN(\mathbf{p}, \mathbf{q}, t)$

$\qquad\qquad + \tfrac{1}{2} \int V(\mathbf{q}_{12}) \, dN(\mathbf{p}_1, \mathbf{q}_1, t) \, dN(\mathbf{p}_2, \mathbf{q}_2, t),$

where $\mathbf{u}$ is the (macroscopic) velocity of the drop of fluid, given by

(7.2.7) $\qquad\qquad\qquad \mathbf{u} = E\{\mathbf{P}\}/E\{M\}$

and $F(.)$ and $V(.)$ are the force and mutual potential between any two molecules. For a fluid not subject to any external force, momentum has no significance, and the properties of the fluid are determined essentially by the local density, pressure and local energy. In the case of a rare gas, most of the contributions to the pressure and local energy arise from the first term on the right-hand side of (7.2.5) and (7.2.6), while in the case of a very dense fluid, significant contributions arise from the second term on the right-hand side of (7.2.5) and (7.2.6). The local density $\rho(\mathbf{r}, t)$ (where $\mathbf{r}$ is defined with macroscopic accuracy) can be defined by

(7.2.8) $\qquad\qquad \rho(\mathbf{r}, t) \, d\mathbf{r} = \int m \, dN(\mathbf{p}, \mathbf{q}, t),$

where the domain of $\mathbf{q}$ is restricted to a sphere with centre $\mathbf{r}$ and radius $|d\mathbf{r}|$.

If we confine our attention to a gas which is not dense and not subject to any external force, we notice that the mean value of $\mathbf{P}$ vanishes, so that the local energy can be written as

(7.2.9) $\qquad\qquad Q(\mathbf{r}, t) = \int \dfrac{p^2}{2m} \, dN(\mathbf{p}, \mathbf{q}, t).$

We next impose the physical condition that the random variables $\{dN(\mathbf{p}, \mathbf{q}, t)\}$ are uncorrelated for disjoint $d\Omega$'s. This assumption is the same as that imposed by Khintchine. The characteristic function of Q defined by

---

[*] This does not mean that all the results are independent of the type of statistics obeyed by the particles. In fact, the utility of multiple point product densities will be discussed in section **7.3**.

(7.2.10) $$\Phi(\theta) = \mathrm{E}\{\exp i\, Q\theta\}$$

can be calculated using the statistical independence of the random variables $dN(\mathbf{p}, \mathbf{q}, t)$, and we thus obtain

(7.2.11) $\quad \Phi(\theta) = \exp\{-\int f_1(\mathbf{p}, \mathbf{q}, t)\}\{1 - \exp(i\theta \mathbf{p}^2/2m)\}d\Omega,$

where $f_1(\mathbf{p}, \mathbf{q}, t)$ is defined by (7.2.2). From (7.2.11) we obtain

(7.2.12) $\quad \mathrm{E}\{Q^2\} - [\mathrm{E}\{Q\}]^2 = \int f_1(\mathbf{p}, \mathbf{q}, t)\left(\frac{p^2}{2m}\right)^2 d\Omega.$

Observing that the right-hand side is of order $\bar{N}$, where $\bar{N}$ is the mean number of molecules in a macroscopic drop of fluid, we find that

(7.2.13) $$\frac{\mathrm{E}\{Q^2\} - [\mathrm{E}\{Q\}]^2}{[\mathrm{E}\{Q\}]^2} \cong \frac{1}{\bar{N}},$$

a result demonstrating the stability of the mean value of Q.

The above derivation of the characteristic function of $Q(\mathbf{r}, t)$ is not valid in the case of moderately dense fluids since the random variables $\{dN(\mathbf{p}, \mathbf{q}, t)\}$ are no longer uncorrelated for disjoint $d\Omega$'s. However, it is possible to calculate the moments of $Q(\mathbf{r}, t)$ as well as other macroscropic quantities. The first two moments of $Q(\mathbf{r}, t)$ can be calculated with the help of the product-density technique and are given by

(7.2.14) $\quad \mathrm{E}[Q(\mathbf{r}, t)] = \int \frac{p^2}{2m} f_1(\mathbf{p}, \mathbf{q}, t)\, d\Omega$

(7.2.15) $\quad \mathrm{E}[Q^2(\mathbf{r}, t)]$

$$= \int \left(\frac{p^2}{2m}\right)^2 f(\mathbf{p}, \mathbf{q}, t)\, d\Omega + \iint \frac{p_1^2 p_2^2}{4m^2} f_2(\mathbf{p}_1, \mathbf{q}_1; \mathbf{p}_2, \mathbf{q}_2; t)\, d\Omega_1\, d\Omega_2,$$

where

(7.2.16) $\quad f_2(\mathbf{p}_1, \mathbf{q}_1; \mathbf{p}_2, \mathbf{q}_2; t)\, d\Omega_1\, d\Omega_2 = \mathrm{E}[dN(\mathbf{p}_1, \mathbf{q}_1; t)\, dN(\mathbf{p}_2, \mathbf{q}_2; t)].$

We next observe that the partial dependence of the random variables $\{dN(\mathbf{p}, \mathbf{q}, t)\}$ arising from the short range correlation of the molecules characteristic of a dense fluid, can be expressed as

(7.2.17) $\quad f_2(\mathbf{p}_1, \mathbf{q}_1; \mathbf{p}_2, \mathbf{q}_2; t) = f_1(\mathbf{p}_1, \mathbf{q}_1; t) f_1(\mathbf{p}_2, \mathbf{q}_2; t)\{1 + g(\mathbf{q}_1, \mathbf{q}_2)\},$

where $g(\mathbf{q}_1, \mathbf{q}_2)$ vanishes everywhere except for those configurations in which $|\mathbf{q}_1 - \mathbf{q}_2|$ is of the order of a few molecular diameters. Using (7.2.17), it is easy to establish that (7.2.15) and (7.2.14) imply (7.2.13). In a similar way it follows that

(7.2.18) $\quad \dfrac{\mathrm{E}\{[Q(\mathbf{r}, t)]^p\} - [\mathrm{E}\{Q(\mathbf{r}, t)\}]^p}{[\mathrm{E}\{Q(\mathbf{r}, t)\}]^p} \cong \dfrac{1}{\bar{N}}, \quad p = 3, 4, 5, \dots.$

It is not difficult to prove relations of the type (7.2.13) and (7.2.18) for Q defined by (7.2.6), and exactly similar considerations apply to other macroscopic quantities. This demonstration goes to show that the mean values of macroscopic quantities like $Q(\mathbf{r}, t)$ are stable against statistical fluctuations.

7.3 Non-equilibrium Statistical Physics

So far we have not attempted to derive any structure for the product densities defined over the phase-space. In fact, if we consider the time-variation of the product densities, the problem bears some analogy to the equations satisfied by population point processes discussed in Chapter 2. This useful analogy, although noted by Nordsieck, Lamb and Uhlenbeck (1940), has, surprisingly enough, not attracted much attention and has had no impact on the literature relating to the fluctuation theorem. To derive the equations satisfied by the product densities, we shall assume that some kind of coarse-grained structure exists in time, and that the product densities have been obtained by an average over the time-parameter in the range $(t - \tau, t + \tau)$ where t is measured on the kinetic time-scale (10^{-9} sec), while τ is of the order of 10^{-12} sec (initial time) (see, for example, Cohen (1962)).

To derive the equation staisfied by $f_1(\mathbf{p}, \mathbf{q}; t)$, we make use of the mean number interpretation as explained in section 7.2, and increase t to $t + \Delta$. We can estimate the time-rate of change in $f_1(\mathbf{p}, \mathbf{q}; t)$ using classical mechanical principles; this, in turn, equals the mean number of particles that are pushed into the phase space $d\Omega$ as a result of the collisions, and naturally this term will be a linear functional of the product density of degree two. Thus, we find, in the absence of external forces, that

$$(7.3.1) \quad \frac{\partial f_1(\mathbf{p}, \mathbf{q}; t)}{\partial t} + \left(\frac{p_i}{m} \right) \frac{\partial f_1(\mathbf{p}, \mathbf{q}; t)}{\partial q_i} = L_c(f_2)$$

$$(7.3.2) \quad L_c(f_2)$$
$$= \int \int [f_2(\mathbf{p} - \Delta\mathbf{p}, \mathbf{q}; \mathbf{p}_1, \mathbf{q} + \sigma\mathbf{k}; t) - f_2(\mathbf{p}, \mathbf{q}; \mathbf{p}_1, \mathbf{q} - \sigma\mathbf{k}; t)] vb\,db\,d\epsilon\,d\mathbf{p}_1,$$

where σ is the diameter of a molecule and $\mathbf{k}$ the unit vector along the line of centres.

The above equation is obtained on the assumption that the change in momentum of the molecules occurs due to collisions only, and this is certainly the case for a rare gas. However, for a dense fluid or liquid, it is necessary to take into account the infrequent nature of the collisions and the change in momentum of the molecules due to the Brownian type of oscillation suffered by the individual molecules. In this case we obtain equation (7.3.1) with the right-hand side replaced by $L_c + L_s$, where

$$(7.3.3) \quad L_s = -(2\tau)^{-1} \int_{-\tau}^{\tau} \int \int F_i^{12} \frac{\partial f_2}{\partial p_i} (\mathbf{p}, \mathbf{q}; \mathbf{p}_1, \mathbf{q}_1; t + s)\,ds\,d\mathbf{p}_1\,d\mathbf{q}_1,$$

where f_2 is the product density, defined "finely" on the initial time-scale, and $\mathbf{F}^{12}$ is the intermolecular force. For a liquid, L_s yields a significant contribution, while for a rare gas L_c is significant. Thus, for a gas, if we make the Poisson approximation

(7.3.4) $f_2(\mathbf{p}_1, \mathbf{q}_1; \mathbf{p}_2, \mathbf{q}_2; t) = f_1(\mathbf{p}_1, \mathbf{q}_1; t) f_1(\mathbf{p}_2, \mathbf{q}_2; t),$

equation (7.3.1) reduces to the familiar Boltzmann equation.

In general, f_1 cannot be obtained unless we seek information about f_2. The function f_2 satisfies an equation which is very similar to (7.3.1). Thus we are led to a hierarchy whose structure, though similar to the $B - B - G - K - Y$ hierarchy (see, for example, Bogoliubov (1962)), has an entirely different probabilistic interpretation. If, however, wall effects are neglected, the two systems become identical in the limit as N tends to infinity.

To deduce the macroscopic properties, we impose the *approximate* independence of the random variables $dN(p, q; t)$. This, in turn, can be expressed by the condition[*]

(7.3.5) $f_n(\mathbf{p}_1, \mathbf{q}_1; \mathbf{p}_2, \mathbf{q}_2, \ldots, \mathbf{p}_n, \mathbf{q}_n; t)$

$$= (1 + \alpha_n) f_1(\mathbf{p}_1, \mathbf{q}_1; t) f_1(\mathbf{p}_2, \mathbf{q}_2; t) \ldots f_1(\mathbf{p}_n, \mathbf{q}_n; t),$$

where α_n is a function of the microscopic coordinates $\mathbf{q}_1, \mathbf{q}_2, \ldots, \mathbf{q}_n$ and a functional of f_1, α_n being negligibly small for large values of $|q_i - q_j|$ $(i, j = 1, 2, \ldots, n, i \neq j)$. This is a physical assumption expressing macroscopic stability as well as the non-linear nature of the process, at least on the kinetic time-scale.

The macroscopic hydrodynamic equations are obtained from (7.3.2) by the use of equations (7.2.3) through (7.2.6). The rate of change of average values of M, $\mathbf{P}$, $\mathcal{P}_{ij}$ and Q is obtained if we integrate both sides of (7.3.2) after multiplication by suitable weight factors. It is to be noted that the time-parameter is measured on the hydrodynamic scale, and as such the evaluation of the right-hand side of (7.3.2) must be done in a careful manner similar to that usually adopted in passing to a coarse-grained structure from the molecular level. If this is done, we obtain the usual Navier–Stokes equation of motion for the momentum components of a drop of fluid, along with the equation of continuity (see, for example, Green (1952) and Srinivasan (1967, 1968)). In view of the non-vanishing nature of the contribution arising from $L_c(f_2)$, the equations will contain the average value of $\mathcal{P}_{ij}$, which in turn is evaluated to first order and related to the rate of strain by means of a transport coefficient. For a compressible gas, there is an additional equation for

---

[*] There are several ways of expressing the weak dependence. There are strong reasons to believe that f_2 must be determined independently. We have chosen a relation of of the type (7.3.5) for purposes of illustration only.

the mean value of Q, which in turn yields the equation of state. Thus the equations represent the mean motion suffered by an elemental drop of fluid. An important question therefore arises at this stage as to whether the mean motion is representative of the true or actual motion of the drop of fluid. That the answer is in the affirmative is a consequence of equation (7.3.5). It also follows that the transport coefficients do not undergo any fluctuation since they are defined as being the ratio of a certain flux to an appropriate force, the fluxes and forces themselves being free from fluctuations. From the current viewpoint, it is clear that turbulent motion is the natural type of motion, and that laminar motion is possible only under special circumstances. At present, equation (7.3.5) is perhaps the only way of expressing such a possibility.

Instabilities of the state of fluid, be it in motion or state of rest, naturally arise from the breakdown of relations of the type (7.3.5). Observed experimental evidence for turbulent motion in certain ranges of Reynolds number can be effectively incorporated in relation (7.3.5) by choosing the norm of the functionals α_n to be comparable to unity.

We also wish to observe that the present approach could be the correct one for the formulation of other problems like the phase transition from gas to liquid. In such a case, it may be necessary to derive an equation satisfied by the second-order product density similar to (7.3.2), and deduce the behaviour of α_2 as defined by (7.3.4). This may, perhaps, provide a sound probabilistic basis for the phenomenological equation proposed by Ornstein and Zernike (1914).

We hope that the method described in this chapter may also be useful in other contexts such as the formation of shock fronts in hydrodynamics and plasmas. The solution to the problem consists in demonstrating the existence of fluctuations (at macroscopic scale) through the approximate solution of the equation satisfied by the second-order product density. A beginning has already been made by Ichimaru, who has proposed kinetic equations for turbulent plasmas on the basis of equations that are satisfied by product densities of the first two orders by using a closure approximation similar to (7.3.5) for $n \geqslant 3$. Ichimaru has estimated the second-order product density and related it to the fluctuation of the energy spectrum of the plasma.

It might be profitable to study the product density of second order in the approximation used by Ichimaru (1970) for a classical turbulent fluid and to characterize α_2 as defined by (7.3.5). Such an approach may provide a quantitative deductive theory of turbulence.

The theory of turbulence has been studied from many angles since Heisenberg and Kolmogorov derived the turbulent velocity spectrum. In almost all these contributions the validity of the Navier–Stokes equation in the turbulent regime is taken for granted. Although many conflicting results have been derived, the basic problem of the transition of flow resulting in non-deterministic characteristics has not attracted much attention (see, for example, Beran (1968)). Landau (1965) conjectured — see also Uhlenbeck (1968) — that the non-linear nature of the Navier—Stokes equation generates different modes in the wave-number space, the number of modes becoming so large that the motion ultimately becomes turbulent. While such an explanation is plausible, it leads to the larger question as to whether non-linearity of any phenomenon necessarily leads to non-determinism. Stemming from these considerations, Srinivasan (1967, 1968) has attempted to study the validity of the basic equations. It has been shown by the author that while turbulent flow satisfies some kind of generalized Navier—Stokes equation of the type anticipated by Reynolds (1895), the momentum equation, at least in the neighbourhood of the transition region, contains a random force term. Also it turns out that the transport coefficients are stochastic functionals of the forces and fluxes. It may be profitable to look for the detailed structure of the force term as well as of the random functional.

7.4 Quantum statistics

In this section we shall discuss a dynamical method of arriving at the equilibrium distribution function corresponding to Maxwell—Boltzmann, Fermi—Dirac, Bose—Einstein and intermediate statistics. The distribution functions corresponding to the first three cases were derived by Moyal (1949) on the assumption that the energy levels over which the particles are distributed are discrete. However, the E-space or velocity space over which the particles are distributed is a continuum, and product-density formalism is a very useful approach in this context[*]. We propose to study the distribution function using multiple product densities. This is best done by setting up equations for the product densities defined in the phase space. Thus the equilibrium conditions combined with the principle of microscopic reversibility lead to the energy spectrum, provided we translate the statistics of particles into probabilities governing the collisions.

Let $dN(\mathbf{v}, t)$ be the number of particles with velocities in the range $(\mathbf{v}, \mathbf{v} + d\mathbf{v})$ at time t. The changes in the distribution are caused by binary collisions in which two particles with velocities in the ranges $(\mathbf{u}, \mathbf{u} + d\mathbf{u})$, $(\mathbf{v}, \mathbf{v} + d\mathbf{v})$ acquire velocities in the ranges $(\mathbf{r}, \mathbf{r} + d\mathbf{r})$, $(\mathbf{s}, \mathbf{s} + d\mathbf{s})$ with probability $\lambda_{\mathbf{uv,rs}} \, d\mathbf{r} \, d\mathbf{s}$. If the particles obey Maxwell—Boltzmann statistics, there is no restriction on the number of particles for the occupancy of each velocity range $d\mathbf{v}$. Thus the occupation numbers n_1 can take all positive integral values in the case of Maxwell—Boltzmann statistics and hence are composed of multiple points in $\mathbf{v}$-space of *all* orders:

---

[*] In fact, Moyal himself pointed out, in one of the discussions (see Kendall (1949)), that the product-density approach would be more appropriate in discussing the case when the energy spectrum is continuous.

$$dN(\mathbf{v}, t) = \sum_{j=1}^{\infty} j \, dN_j(\mathbf{v}, t).$$

Thus the probability of collision that carries over the two particles – one of which is in the range $(\mathbf{v}, \mathbf{v} + d\mathbf{v})$, the other being in $(\mathbf{u}, \mathbf{u} + d\mathbf{u})$ – into the ranges $(\mathbf{s}, \mathbf{s} + d\mathbf{s})$, $(\mathbf{r}, \mathbf{r} + d\mathbf{r})$, is proportional to the product of $dN(\mathbf{v}, t)$ and $dN(\mathbf{u}, t)$. In the case of Bose–Einstein statistics, any number of particles can occupy a particular range $d\mathbf{v}$, as in the case of Maxwell–Boltzmann particles. However, the collision probability will depend on the number of particles that occupy the ranges $(\mathbf{r}, \mathbf{r} + d\mathbf{r})$ and $(\mathbf{s}, \mathbf{s} + d\mathbf{s})$.

If, on the other hand, the particles obey Fermi statistics, $dN(\mathbf{v}, t)$ can take the values 0 and 1 only, and hence mean values of $dN(\mathbf{v}, t)$ and their products will lead to simple product densities. Apart from this, the collision probability will also contain additional factors, ensuring that in a binary collision the particles acquire velocities only in those ranges which are not already occupied. We now proceed to write down the dynamical equations.

7.4.1   Maxwell–Boltzmann statistics

In this case, it is unnecessary to deal with the random variable $dN_i(\mathbf{v}, t)$; it is sufficient to study the changes in the expectation value of $dN(\mathbf{v}, t)$ corresponding to the total number of particles. Taking into account the changes that occur due to collisions in the time-interval $(t, t + \Delta)$, we find that

$$(7.4.1) \quad \frac{\partial}{\partial t} E\{dN(\mathbf{v}, t)\}$$

$$= -\int_{\mathbf{u}} \int_{\mathbf{r}} \int_{\mathbf{s}} E\{dN(\mathbf{v}, t) \, dN(\mathbf{u}, t)\} \, \lambda_{\mathbf{uv,rs}} \, d\mathbf{r} \, d\mathbf{s} + \int_{\mathbf{u}} \int_{\mathbf{r}} \int_{\mathbf{s}} E\{dN(\mathbf{r}, t) \, dN(\mathbf{s}, t)\} \lambda_{\mathbf{rs,uv}} \, d\mathbf{u}$$

or

$$\frac{\partial f_1(\mathbf{v}, t)}{\partial t}$$

$$= -\int \int \int f_2(\mathbf{u}, \mathbf{v}, t) \, \lambda_{\mathbf{uv,rs}} \, d\mathbf{u} \, d\mathbf{r} \, d\mathbf{s} + \int \int \int f_2(\mathbf{r}, \mathbf{s}, t) \, \lambda_{\mathbf{rs,uv}} \, d\mathbf{u} \, d\mathbf{r} \, d\mathbf{s}.$$

In a similar way, we can write equations expressing the rate of change of $f_2(.,., t)$ in terms of third-order product densities, and so on. However, molecular chaos, or complete *stosszahlansatz*, implies statistical independence of the random variables $dN(.,., t)$. This is expressed by

$$(7.4.2) \quad f_m(\mathbf{v}_1, \mathbf{v}_2, \ldots, \mathbf{v}_m, t) = f_1(\mathbf{v}_1, t) f_1(\mathbf{v}_2, t) \ldots f_1(\mathbf{v}_m, t).$$

In addition, the principle of microscopic reversibility implies that

$$(7.4.3) \qquad\qquad \lambda_{\mathbf{uv,rs}} = \lambda_{\mathbf{rs,uv}}.$$

The equilibrium configuration is obtained by setting $\partial f_1/\partial t$ equal to zero. These conditions yield the functional equation for f_1:

$(7.4.4)$ $$f_1(\mathbf{r})f_1(\mathbf{s}) = f_1(\mathbf{v})f_1(\mathbf{u}).$$

Imposition of energy conservation at each collision yields

$(7.4.5)$ $$f_1(\mathbf{v}, t) = A\,e^{-\beta v^2},$$

where A and β can be identified with the familiar constants of the Boltzmann distribution.

7.4.2 Fermi–Dirac and Bose–Einstein statistics

In a similar fashion, we can write the differential equation for product density $f_1(u, t)$ in the case of Fermi–Dirac particles, assuming that $dN(u, t)$ can only be equal to $dN_1(u, t)$, which corresponds to the occupancy by 1 or 0 particles only of the du region in the velocity space. To accord with the fact that the presence of a particle in a region $d\mathbf{r}$ inhibits other particles from going into it, the density of states is represented by $\left[1 - \dfrac{dN(\mathbf{r}, t)}{d\mathbf{r}}\right]d\mathbf{r}$. By the same token, for a Bose gas with unlimited occupancy of each velocity range to accord with the fact that the existence of particles in each region facilitates other particles going into the region, the increased density of states can be represented by $\left[1 + \dfrac{dN(\mathbf{r}, t)}{d\mathbf{r}}\right]d\mathbf{r}$. For Fermi–Dirac particles we obtain

$(7.4.6)$ $\dfrac{\partial}{\partial t}\,\mathrm{E}\{dN_1(\mathbf{v}, t)\}$

$= \int \mathrm{E}\{dN_1(\mathbf{v}, t)\,dN_1(\mathbf{u}, t)\,[d\mathbf{r} - dN_1(\mathbf{r}, t)][d\mathbf{s} - dN_1(\mathbf{s}, t)]\}\,\lambda_{\mathbf{uv},\,\mathbf{rs}}$

$\quad + \int \mathrm{E}\{dN_1(\mathbf{r}, t)\,dN_1(\mathbf{s}, t)\,[d\mathbf{v} - dN_1(\mathbf{v}, t)][d\mathbf{u} - dN_1(\mathbf{u}, t)]\}\,\lambda_{\mathbf{rs},\,\mathbf{uv}}.$

Proceeding exactly as in the case of Maxwell–Boltzmann particles, we find that

$(7.4.7)$ $f_1(\mathbf{v}, t)f_1(\mathbf{u}, t)[1 - f_1(\mathbf{r}, t) - f_1(\mathbf{s}, t)]$

$\qquad = f_1(\mathbf{r}, t)f_1(\mathbf{s}, t)[1 - f_1(\mathbf{v}, t) - f_1(\mathbf{u}, t)].$

If we impose energy conservation in each collision and take into account the total energy and particle number of the system, we arrive at

$(7.4.8)$ $$f_1(\mathbf{v}, t) = \frac{1}{A\,e^{\beta v^2} + 1},$$

which can be identified with the Fermi distribution by a proper choice of A and β.

In an exactly similar manner, we obtain the energy spectrum of the Bose particles, if we use the appropriate density of states.

7.4.3 Intermediate statistics

We next take the case of particles whose occupancy in any region of velocity space is restricted to 0 or 1 or 2. The density of states is taken to be of the Bose type. Hence the total number in each infinitesimal range $d\mathbf{v}$ is given by

$$(7.4.9) \qquad dN(\mathbf{v}, t) = dN_1(\mathbf{v}, t) + 2dN_2(\mathbf{v}, t).$$

The equations for the rate of change of $dN_1(\mathbf{v}, t)$ and $dN_2(\mathbf{v}, t)$ are given by

$$(7.4.10a) \quad \frac{\partial}{\partial t} \mathrm{E}\{dN_1(\mathbf{v}, t)\}$$

$$= -\int_\mathbf{u}\int_\mathbf{r}\int_\mathbf{s} \mathrm{E}\left\{\sum_i i\, dN_i(\mathbf{u}, t)\, dN_1(\mathbf{v}, t)\, d\alpha(\mathbf{r}, t)\, d\alpha(\mathbf{s}, t)\right\} \lambda_{\mathbf{uv}, \mathbf{rs}}$$

$$+ \int_\mathbf{u}\int_\mathbf{r}\int_\mathbf{s} \mathrm{E}\left\{\sum_i 2i\, dN_2(\mathbf{u}, t)\, dN_i(\mathbf{v}, t)\, d\alpha(\mathbf{r}, t)\, d\alpha(\mathbf{s}, t)\right\} \lambda_{\mathbf{uv}, \mathbf{rs}}$$

$$+ \int_\mathbf{u}\int_\mathbf{r}\int_\mathbf{s} \mathrm{E}\left\{\sum_{ij} ij\, dN_i(\mathbf{r}, t)\, dN_j(\mathbf{s}, t)\, d\alpha(\mathbf{u}, t)\, d\alpha(\mathbf{v}, t)\right\} \lambda_{\mathbf{rs}, \mathbf{uv}}$$

$$(7.4.10b) \quad \frac{\partial}{\partial t} \mathrm{E}\{dN_2(\mathbf{v}, t)\}$$

$$= -2\int_\mathbf{u}\int_\mathbf{r}\int_\mathbf{s} \mathrm{E}\left\{\sum_i dN_i(\mathbf{u}, t)\, dN_2(\mathbf{v}, t)\, d\alpha(\mathbf{r}, t)\, d\alpha(\mathbf{s}, t)\right\} \lambda_{\mathbf{uv}, \mathbf{rs}}$$

$$+ \int_\mathbf{u}\int_\mathbf{r}\int_\mathbf{s} \mathrm{E}\left\{\sum_{ij} ij\, dN_i(\mathbf{r}, t)\, dN_j(\mathbf{s}, t)\, d\alpha(\mathbf{u}, t)\left[1 + \frac{dN_1(\mathbf{v}, t)}{d\mathbf{v}}\right]\right.$$

$$\times \left.\left[1 - \frac{dN_2(\mathbf{v}, t)}{d\mathbf{v}}\right]\right\} \lambda_{\mathbf{uv}, \mathbf{rs}}\, d\mathbf{v},$$

where

$$(7.4.11) \quad d\alpha(\mathbf{u}, t) = d\mathbf{u}\left[1 + \sum_i \frac{dN_i(\mathbf{u}, t)}{d\mathbf{u}}\right]\left[1 - \frac{dN_2(\mathbf{u}, t)}{d\mathbf{u}}\right]$$

and the factors $\left[1 - \dfrac{dN_2}{d\mathbf{u}}\right]$ have been used to indicate that the total occupancy is limited to 2. Imposing microscopic reversibility, we obtain

$$(7.4.12) \quad \frac{\partial}{\partial t} \mathrm{E}\{dN(\mathbf{v}, t)\} = \int_\mathbf{r}\int_\mathbf{s}\int_\mathbf{u} \mathrm{E}\left\{dN(\mathbf{r}, t)\, dN(\mathbf{s}, t)\, \lambda_{\mathbf{uv}, \mathbf{rs}}\, d\mathbf{u}\, d\mathbf{v}\right.$$

$$\times \left.\left[1 + \frac{dN_1(\mathbf{v}, t)}{d\mathbf{u}} - \frac{dN_2(\mathbf{v}, t)}{d\mathbf{v}}\right]\left[1 + \frac{dN_1(\mathbf{u}, t)}{d\mathbf{u}} - \frac{dN_2(\mathbf{u}, t)}{d\mathbf{u}}\right]\right\}$$

$$- \int_\mathbf{r}\int_\mathbf{s}\int_\mathbf{u} \mathrm{E}\left\{dN(\mathbf{u}, t)\, dN(\mathbf{v}, t)\, \lambda_{\mathbf{uv}, \mathbf{rs}}\, d\mathbf{s}\, d\mathbf{r}\, \times\right.$$

$$\times \left[1 + \frac{dN_1(\mathbf{r}, t)}{d\mathbf{r}} - \frac{dN_2(\mathbf{r}, t)}{d\mathbf{r}}\right]\left[1 + \frac{dN_1(\mathbf{s}, t)}{d\mathbf{s}} - \frac{dN_2(\mathbf{s}, t)}{d\mathbf{s}}\right]\right\}.$$

At the equilibrium configuration, the left-hand side of (7.4.12) is zero. Proceeding as before, we obtain

$$(7.4.13) \quad f(\mathbf{r}, t) f(\mathbf{s}, t)[1 + f^1(\mathbf{v}, t) - f^2(\mathbf{v}, t)][1 + f^1(\mathbf{u}, t) - f^2(\mathbf{u}, t)]$$
$$= f(\mathbf{u}, t) f(\mathbf{v}, t)[1 + f^1(\mathbf{r}, t) - f^2(\mathbf{r}, t)][1 + f^1(\mathbf{s}, t) - f^2(\mathbf{s}, t)].$$

Taking into account energy conservation in each collision, we see that the above equation can be satisfied if

$$(7.4.14) \qquad\qquad f^1(\mathbf{v}, t) = \frac{A e^{+\beta v^2}}{1 + A e^{+\beta v^2} + A^2 e^{+2\beta v^2}}$$

and

$$(7.4.15) \qquad\qquad f^2(\mathbf{v}, t) = \frac{1}{1 + A e^{+\beta v^2} + A^2 e^{+2\beta v^2}}.$$

Thus the total product density is given by

$$(7.4.16) \qquad\qquad f(\mathbf{v}, t) = \frac{A e^{+\beta v^2} + 2}{1 + A e^{+\beta v^2} + A^2 e^{+2\beta v^2}}$$

which is in accordance with the distribution for the intermediate statistics or Gentile statistics (see ter Haar (1952)), the highest occupancy for each state being only 2.

The above result can be generalized to the case when the maximum number of particles that can occupy a state is p. If we let p tend to infinity, we recover Bose–Einstein statistics.

8   MANAGEMENT SCIENCE

8.1   Introduction

The object of this chapter is to highlight the applications of point processes to some of the problems of management science. The theory of queues is fairly old and well established, and innumerable contributions have been made since the beginning of this century. However, its connection with the theory of point processes is of fairly recent origin and is mainly due to D.G. Kendall (1951, 1964) and Khintchine (1955). Feller (1966) focussed the attention of workers in this field on the need of a general unified technique to handle problems of queues and inventories. Again, the theory of inventories has been investigated from many angles, particularly from the viewpoint of applications. The origin of the theory can be traced back to certain models of dams and water-storage facilities developed by Moran (1954, 1959) and Prabhu (1964, 1965). In this chapter we shall attempt to fulfil the task envisaged by Feller of identifying the renewal processes imbedded in the stochastic processes of queues and inventories. A similar viewpoint will be adopted for the study of problems of reliability.

8.2   Single server queuing theory

Let us consider a waiting system where the population of customers is infinite and customers arrive individually, there being no balking, reneging, or priorities in service. The epochs of arrivals form a renewal process, so that the time-intervals between successive arrivals are independently and identically distributed, the common density function being $f(.)$. Queue discipline is assumed to be strictly adhered to, so that service is in order of arrival. It is also assumed that there is only one service facility and that the service times of customers are independently and identically distributed, the common density function being $\phi(.)$. The queuing process thus generated on the waiting system can be viewed as a positive, integral-valued stochastic process $L(.)$, where $L(t)$ denotes the queue length or the number of individuals or items in the queue at epoch t, including the one that is being serviced. The points of discontinuity of $L(.)$ denote the epochs at which an arrival or a departure occurs. Thus a departure corresponds to a downward jump of $L(.)$ while arrival corresponds to an upward jump, the magnitude of the jump being always unity.

The stochastic process $L(.)$ is non-Markovian unless both $f(.)$ and $\phi(.)$

are exponential functions of their arguments. If

$$(8.2.1) \qquad f(x) = \lambda e^{-\lambda x} \quad \lambda > 0$$

$$(8.2.2) \qquad \phi(x) = \mu e^{-\mu x} \quad \mu > 0,$$

then it is easy to show that the function $\pi_n(.)$ defined by

$$(8.2.3) \qquad \pi_n(t) = \text{Pr}\,\{L(t) = n\}$$

satisfies

$$(8.2.4) \quad \pi'_n(t) = -(\lambda + \mu)\pi_n(t) + \lambda\pi_{n-1}(t) + \mu\pi_{n+1}(t) \quad n \geqslant 1$$

and

$$(8.2.5) \qquad \pi'_0(t) = -\lambda\pi_0(t) + \mu\pi_1(t).$$

The above set of equations can be solved by the generating function technique (see, for example, Saaty (1961), Prabhu (1965)). If, however, either of the conditions (8.2.1) or (8.2.2) is relaxed, the Markovian property of the process $L(.)$ is lost and it is not possible to arrive at differential equations satisfied by the $\pi_n(.)$.

To study the queuing process for general $f(.)$ and $\phi(.)$, two courses of action are open. The first consists in attacking the problem directly by dealing with a vector process

$$Z(.) = (L(.), A(.), D(.)),$$

where $A(t)$ $(D(t))$ is the time that has elapsed since the last arrival (departure) in the queue, measured from t. Obviously the vector process $Z(.)$ is Markovian in nature, and it may be possible to derive differential or integral equations for the joint probability density function of $L(.)$, $A(.)$ and $D(.)$ (see problems 8.2 and 8.3, below). The resulting equations are very inconvenient, if not intractable. The second course of action consists in the introduction of the random variable $W(.)$, representing the so-called *virtual waiting time* (i.e. the time an individual has to wait for the commencement of his service were he to arrive at epoch t). The quantity $W(t)$ denotes the cumulative time taken to render service to all the customers present at epoch t. In other words, $W(t)$ is the sum of $L(t)$ random variables, $L(t) - 1$ of which are distributed with common density function $\phi(.)$, the $L(t)$th random variable (corresponding to the individual being served at epoch t) representing the residual service-time. In this approach, queue length is converted into the effective load on the counter. The random variable is also known in the literature as the *load on the counter*. It is this approach that has been successfully employed for the study of the queuing process (see, for example, Prabhu (1965), Cohen (1969)). The stochastic process $W(.)$ has been studied by identifying it with the sum of a random number of random variables. D.G. Kendall (1951) observed that if the arrivals form a Poisson process (i.e., $f(.)$ is exponential in nature), then $L(t_i)$ is a Markov chain, where t_i is the epoch corresponding to a departure. In other words, the queuing process with Poisson arrivals has a Markov chain imbedded in it. This fact enables us to obtain all the properties of the stochastic process $L(.)$. A full account of the method and its generalization to non-Poisson arrivals can be found in the monograph of Prabhu (1965). A similar approach has been proposed by Takacs (1967) who uses combinatorial methods to arrive at the distribution of $W(.)$.

It is relevant to inquire whether the theory of regenerative events can be used to arrive at the distribution of $W(.)$. In the case of queues with Poisson arrivals, it is well known that any epoch t at which $W(t) = 0$ is *regenerat* (see, for example, D.G. Kendall (1964), Kingman (1964)). In fact, such poin whenever they occur, form a continuum with probability one. These points are members of the interval over which the counter is idle (due to lack of customers) (see also section **3.6**). When the arrivals form a renewal process, Srinivasan and Subramanian (1969) have observed that the epochs of arrival constitute a set of regenerative events, so that $\{W(t_i)\}$ is an *imbedded regenerative stochastic process* (see section **3.6**). Let us define $\pi(x, y, t)$ by

$$(8.2.6) \quad \pi(x, y, t) = \lim_{\Delta, \Delta' \to 0} \Pr\{x \leqslant W(t) < x + \Delta \mid y = W(\Delta') > W(0)\}/\Delta.$$

Following the procedure due to Bellman and Harris (1948), we can obtain the backward integral equation satisfied by $\pi(x, y, t)$ by identifying the first arrival counted from the time-origin as a renewal or regenerative event. To achieve this, it is convenient to note that the following three mutually exclusive events are possible:

(i) no customer arrives up to time t, and this occurs with probability $\int_t^\infty f(u)\, du$;

(ii) a customer arrives at an epoch in $(u, u + du)$ before the counter is free (the load drops to zero), and this occurs with probability $f(u)\, d$ $(u < \min(y, t))$;

(iii) a customer arrives at an epoch in $(u, u + du)$ after the counter has freed itself of the initial load y, and this occurs with probability $f(u)\, du$ $(y < u < t)$.

We then observe that in event (ii), the contribution to $\pi(x, y, t)$ can be written as

$$\lim_{\Delta_1, \Delta_2 \to 0} \int_0^{\min(y, t)} f(u)\, du \int_{v=0}^\infty \Pr\{x \leqslant W(t) \leqslant x + \Delta_1 \mid y = W(\Delta_2) > W(0),$$

$$(8.2.7) \qquad y - u + v = W(u + du) > W(u)\} \phi(v)\, dv/\Delta_1,$$

which, by virtue of the t-homogeneous property of the queuing process and the regenerative (renewal) nature of the epoch u, can be written as

$$(8.2.8) \quad \lim_{\Delta \to 0} \int_0^{\min(y,t)} f(u)\, du \int_{v=0}^{\infty} \Pr\{x \leqslant W(t-u) \leqslant x + \Delta \mid y - u + v$$

$$= W(u + du) > W(u)\}\, \phi(v)\, dv/\Delta$$

$$= \int_0^{\min(y,t)} f(u)\, du \int_0^{\infty} dv\, \phi(v)\, \pi(x, y - u + v, t - u).$$

In a similar manner, in event (iii), the contribution to $\pi(x, y, t)$ can be written as

$$(8.2.9) \quad \lim_{\substack{\Delta_1 \to 0 \\ \Delta_2 \to 0}} \int_y^t f(u)\, du \int_{v=0}^{\infty} \Pr\{x < W(t) < x + \Delta_1 \mid y = W(\Delta_2) > W(0),$$

$$v = W(u + du) > W(u)\}\, \phi(v)\, dv/\Delta_1,$$

which, by virtue of the t-homogeneous nature of the queuing process and the regenerative nature of the epoch u, can be written as

$$\lim_{\Delta \to 0} \int_y^t f(u)\, du \int_0^{\infty} dv \Pr\{x < W(t-u) < x + \Delta \mid v = W(u + du) > W(u)\}\, \phi(v)/\Delta$$

$$(8.2.10) \quad = \int_y^t f(u)\, du \int_0^{\infty} dv\, \phi(v)\, \pi(x, v, t - u).$$

Adding all these contributions, we have

$$(8.2.11) \quad \pi(x, y, t) = \int_t^{\infty} f(u)\, du\, \delta(x - \max(0, y - t))$$

$$+ \int_0^{\infty} dv \int_0^{\min(y,t)} f(u)\, \pi(x, y - u + v, t - u)\, \phi(v)\, du$$

$$+ H(t - y) \int_0^{\infty} dv \int_y^t f(u)\, \pi(x, v, t - u)\, \phi(v)\, du.$$

Introducing the Laplace transform of $\pi(x, y, t)$ by

$$(8.2.12) \quad \pi^*(x, p, s) = \int_0^{\infty} \int_0^{\infty} \pi(x, y, t)\, e^{-yp - st}\, dy\, dt \quad \mathrm{Re}\, s > 0, \quad \mathrm{Re}\, p > 0,$$

we find, after some calculation (see Appendix to this chapter), that

$$(8.2.13) \quad \pi^*(x, p, s)$$

$$= e^{-px} \frac{[1 - f^*(p + s)]}{p + s} + \frac{\delta(x)}{p}\left[\frac{1 - f^*(s)}{s} - \frac{1 - f^*(p + s)}{p + s}\right] + A,$$

where

$$(8.2.14) \quad A = \frac{f^*(p+s)}{2\pi i} \int_{\substack{\sigma-i\infty \\ 0 < \sigma\, \mathrm{Re}\, p}}^{\sigma+i\infty} \frac{\pi^*(x, p', s)\phi^*(-p')\, dp'}{p - p'}$$

$$+ \frac{[f^*(s) - f^*(p+s)]}{2\pi i p} \int_{\sigma-i\infty}^{\sigma+i\infty} \pi^*(x, p', s)\phi^*(-p')\, dp'.$$

The above integral equation can be solved for certain special forms of $\phi(.)$. Srinivasan, Subramanian and Vasudevan (1972) have demonstrated the possibility of solving equations of the type (8.2.13) when $\phi^*(.)$ is a rational function $= \alpha(.)/\beta(.)$, where α is a polynomial of degree n and β another polynomial of degree less than n.

Next we note that the above analysis can be carried out for yet another type of density function $P(x, y, t)$ defined by

$$(8.2.15) \quad P(x, y, t)$$

$$= \lim_{\Delta, \Delta', \Delta'' \to 0} \Pr\{0 < x \leqslant W(t) \leqslant x + \Delta, W(t + \Delta') > W(t) | y = W(\Delta'') > W(0)\}/\Delta$$

The above density function is a product density with respect to t, and the integral of $P(x, y, t)$ over the entire range of x ($x > 0$) can be interpreted as the average intensity of arrivals at any time t when the counter is occupied (busy). It is easy to derive the backward integral equation satisfied by $P(x, y,$ using the same arguments that led to (8.2.11):

$$(8.2.16) \quad P(x, y, t) = f(t)\, \delta(x - \max(0, y - t))$$

$$+ \int_0^\infty dv \int_0^{\min(y, t)} f(u)\, P(x, y - u + v, t - u)\, \phi(v)\, du$$

$$+ H(t - y) \int_0^\infty dv \int_y^t f(u)\, P(x, v, t - u)\, \phi(v)\, du.$$

The Laplace transform solution of $P(x, y, t)$ recurs on parallel lines.

The function $\pi(x, y, t)$ assumes importance viewed from many other angles. Apart from throwing light on the length of queues (see problem 8.4, below), it can also describe a certain counting process associated with the load. To make this point clear, we note that in view of the uniformity of service, we can discuss in an unambiguous manner the random variable $N(x, y, t)$ representing the number of occasions at which $W(t)$ crosses a given level x ($x > 0$) *from above*. In the product-density notation of Ramakrishnan (1950) (see Chapter 2), we have

$$\mathrm{E}\,[d_t N(x, y, t)] = \Pr\{d_t N(x, y, t) = 1\} + o(dt)$$

$$= \int_{x}^{x+c't} \pi(x', y, t) \, dx' + o(dt)$$

(8.2.17) $$= \pi(x, y, t) \, dt + o(dt).$$

Thus $\pi(x, y, t)$ is the product density in t of degree one of x-crossings of the load $W(t)$. Thus the limit of $\pi(x, y, t)$, as x tends to zero *from above*, represents the product density of degree one of the epochs at which the load $W(t)$ drops to zero. The expected number of zero-crossings (droppings) of $W(t)$ over an interval $(0, T)$ is given by

(8.2.18) $$\lim_{x \to 0^+} \int_{0}^{T} E \left[d_t N(x, y, t) \right] = \lim_{x \to 0^+} \int_{0}^{T} \pi(x, y, t) \, dt.$$

It is well known (see Prabhu (1965)) that the zero-crossings of $W(t)$ will recur if

(8.2.19) $$\int_{0}^{\infty} t\phi(t) \, dt / \int_{0}^{\infty} tf(t) \, dt < 1.$$

The above condition ensures that the duration of any busy period (the time-interval in which the load $W(t) > 0$) is finite valued with probability one. In such a case we can deal with the point process generated by the zero-crossings of $W(t)$. The product densities of the zero-crossings can be expressed in terms of $\pi(x, y, t)$ and $P(x, y, t)$. Defining $h_1(y, t)$ by

(8.2.20) $$h_1(y, t) = \lim_{x \to 0^+} \pi(x, y, t),$$

we note that the stationary value of $h_1(y, t)$ (which is non-zero if (8.2.19) is fulfilled) can be obtained. The first-order product density does not characterize the point process since the point process is of the renewal type only if the inputs (arrivals) form a Poisson process. The second-order product density defined by

(8.2.21) $$h_2(y, t_1, t_2) = \lim_{x', x \to 0} E \left[d_{t_1} N(x', y, t_1) \, d_{t_2} N(x, y, t_2) \right] / dt_1 \, dt_2$$

can be readily expressed in terms of $h_1(y, t)$ and $P(x, y, t)$. Using elementary arguments, we find that

(8.2.22) $$h_2(y, t_1, t_2) = \int_{0}^{t_1} dx \int_{0}^{t_1 - x} du \, P(x, y, t_1 - x - u)\phi(u)$$

$$\times \int_{0}^{t_2 - t_1} dv \, f(x + u + v) \int_{0}^{t_2 - t_1 - v} h_1(z, t_2 - t_1 - v)\phi(z) \, dz.$$

By proceeding to the limit as t_1 and t_2 tend to infinity in such a way that $t_2 - t_1 = \tau$, we find that

(8.2.23) $\lim h_2(y, t_1, t_2) = R(\tau)$

$$= \int_0^\infty dx\, P(x, \infty) \int_0^\infty du\, \phi(u) \int_0^\tau dv\, f(x + u + v) \int_0^{\tau - v} h_1(z, \tau - v)\, \phi(z)\, dz,$$

where

(8.2.24) $$P(x, \infty) = \lim_{t \to \infty} P(x, y, t).$$

Hence it is clear that once the function $P(x, y, t)$ is known, all the correlational properties of the queue can be readily obtained. It is also convenient to study the correlational structure of the departures. Daley (1968) studied some aspects of the correlational structure of the output process. He showed that for a stationary queue satisfying the condition (8.2.19) the output process generated by service times obeying an exponential distribution is a renewal process if the arrivals form a Poisson process, in which case the output process is also a Poisson process. It may be worth while to examine the departure pattern through the product-density approach. It is also profitable to study the actual waiting-time suffered by the customers. This is best done by studying the counting process generated by $\mathfrak{N}(x, t)$, the number of customers who depart (after getting served) at epochs before t after waiting for a time less than or equal to x. In such a case we have to study point processe in the product space of x and t. A beginning in this direction has been made by Srinivasan, Subramanian and Vasudevan (1972) who have dealt with the corresponding product densities with the help of the function $\rho(x, y, t)$ defined by (8.2.15). More work in this direction is desirable.

Finally we wish to observe that the method of approach presented in this section is equally applicable to the case when the service-times of the customers form a Markov chain defined by the transition probability $\phi(.|.)$. In such a case, we can define $Q(x, y, z, t)$ by

(8.2.25) $Q(x, y, z, t)$

$= \lim_{\Delta, \Delta', \Delta'' \to 0} \Pr\{x < W(t) < x + \Delta, W(t + \Delta') > W(t) | Z(0) = z,$

$\qquad\qquad W(\Delta'') = y > W(0)\}/\Delta,$

where the random variable $Z(t)$ is the service time of any individual customer arriving at epoch t. Using arguments similar to those used in deriving (8.2.11), we find that

(8.2.26) $Q(x, y, z, t) = f(t)\, \delta(x - \max(0, y - t))$

$\qquad\qquad + \int_0^\infty dv \int_0^{\min(y,t)} du\, f(u)\, Q(x, y - u + v, v, t - u)\, \phi(v|z)$

$\qquad\qquad + H(t - y) \int_0^\infty \phi(v|z)\, dv \int_y^t f(u)\, Q(x, v, v, t - u)\, du, \quad y >$

A similar extension is also possible to cover the case when the time-intervals between successive arrivals also form another Markov chain. An attempt to solve (8.2.26) by Laplace transform technique might well repay the effort.

8.3 Theory of inventories

The theory of inventories deals with storage problems of various commodities, and the problems were first dealt with by econometricians and business administrators. During the last decade several models have been formulated. A systematic study of these models has been made by Arrow, Karlin and Scarf (1958). A more recent account of such models can be found in the monograph of Hadley and Whitin (1963). However, most of the work leans heavily on renewal theory and the random walk, and it may be useful to investigate the applicability of the techniques and results of general point processes. Not much work has been done in this direction and in this section we shall attempt to explain what has so far been achieved.

An inventory is simply a quantity of material stored for the purpose of future sales, distribution, or production. A class of problems arises when the input (receipts) into the stores (or arrivals) and the output from the stores (issues or sales) are non-deterministic in nature. The non-determinism itself arises from the non-deterministic nature of either the epochs of input or of output or of the quantities of input or output; a number of models arise by appropriate combination of these. Let us consider a fairly general case in which the epochs of input and output form two independent stationary point processes. Likewise let the quantities of input (output), *assumed to be discrete in nature* at different epochs, be independently and identically distributed with common probability frequency function $p_{.}(q_{.})$, so that the quantity of input A_t that is added to the stock were there to be an input is determined by the relation

(8.3.1) $$\Pr\{A_t = n\} = \beta_n \quad n = 1, 2, \ldots,$$

where

(8.3.2) $$\sum_{n=1}^{\infty} \beta_n = 1.$$

Similarly the quantity of output B by which the stock would be depleted were there to be an output is determined by the relation

(8.3.3) $$\Pr\{B_t = n\} = \gamma_n \quad n = 1, 2, \ldots,$$

where

(8.3.4) $$\sum_{n=1}^{\infty} \gamma_n = 1.$$

It is clear that the stock level $X(t)$ satisfies the equation

$$(8.3.4) \quad \begin{cases} X(t) = X_0 + \int\limits_0^t A_u \, dN_i(u) - \int\limits_0^t B_u \, dN_0(u) \\[2mm] X(0) = X_0, \end{cases}$$

where $N_i(u)$ and $N_0(u)$ represent the counting process corresponding respectively to the input and output epochs. The above equation is easy to interpret as long as $X(t) > 0$. There is a possibility that the output, which is nothing more than the demand on the stores arriving at a certain epoch, might exceed the stock level at that epoch. In such an eventuality there are two possible courses of action which lead to different models of inventory:

(i) $X(t)$ is allowed to drop to negative integral values. This assumption implies that whenever a demand exceeds the stock, demand is partially satisfied with the available stock, the remainder being supplied on receipt of further stock. This implies that demands occurring during the period in which the warehouse (store) is empty are also satisfied at a later period when a fresh supply of stock (input) is received. This is what is known as *full back-logging of demand orders*. Thus $X(t)$ satisfies (8.3.5) for all values of $X(t)$.

(ii) $X(t)$ is always assumed to be greater than or equal to zero. This implies that if the quantity of demand exceeds the stock, the demand is only partially met. Any demand that occurs during the period when the stock level is zero is lost. This generally happens in a competitive market since the demand can always be satisfied by another competing warehouse. In this, (8.3.5) must be modified to read as

$$(8.3.5a) \quad X(t) = \min \{0, X_0 + \int\limits_0^t A_u \, dN_i(u) - \int\limits_0^t B_u dN_0(u)\}.$$

Most of the investigations relating to inventory models use assumption (i) since the solution of such models does not present any major difficulty. It could be profitable to investigate the random equation (8.3.5a). For instance, in risk theory, which describes actuarial problems, the problem is one of first passage to zero level. Considerable work has been done in this direction by making some simple assumptions about the nature of point processes. Usually it is assumed that point processes are Poisson in nature, and this results in considerable simplification. The reader is referred to the work of Seal (1969).

There are other types of models known as $(S-s)$ models in which quantity of input is related to the stock level. A large number of models can be studied by assuming that $N_0(u)$ is a recurrent point process and $X(0)$ is chosen to be S. To start with, there is no input and the stock level goes on decreasing from the initial value S until it reaches a level less than or equal to s. At the epoch corresponding to the first demand that causes $X(t)$ to drop to a value equal to or lower than s, an order is placed for the quantity $S - s$. Usually this order materializes after a certain length of time, known as "lead time". Of course the aim in ordering $S - s$ is to prevent the inventory from exceeding an upper level S, since the very maintenance of an inventory may of itself be prohibitively costly. Again, two models are possible according as the lead time is deterministic or non-deterministic. In either case there is a possibility that the stock level drops to zero before the arrival of fresh stock of quantity $S - s$. Here again we can follow either of the two courses of action (i) and (ii) explained above. A complete discussion of the S–s inventory models with complete back-logging of orders can be found in the work of Hadley and Whitin (1963). Not much work has been done corresponding to the case when the demands that arise during periods of emptiness of the warehouse are lost. We shall give an example here by discussing a simple S–s model of this type when the demands are Poissonian in nature and the lead times are independently and identically distributed. A model of this type has been discussed recently by Srinivasan and Ramani (1974) with special reference to the first passage to zero stock level.

Let us restate the main assumptions of the model. The epochs of output (demands) are assumed to form a Poisson process with parameter λ, and the quantity of demand is always one unit. When the stock level reaches s, action is taken in the form of an order (from a main supply source) for the quantity $S - s$, which materializes in the form of actual stock in the warehouse after a random time L governed by the density function $\alpha(.)$. The lead times at different epochs are assumed to be independently and identically distributed. In a problem of this type two quantities are of main interest. One is of course the probability frequency function of the stock level $p_n(t)$, defined by

$$(8.3.6) \qquad p_n(t) = \Pr\{X(t) = n \mid X(0) = S\}.$$

The other quantity is the probability density governing the time to the first stock-out. It may also be relevant to deal with the point process generated by the epochs of stock-out. A typical plot of stock level is shown in Fig. 8.1. We next seek a renewal process imbedded in the stochastic process $X(t)$. The epochs at which the stock level materializes form a set of regenerative events in the sense that the value of $X(t + 0)$, if t were to be such an epoch, would result in $\{X(u) (u < t)\}$ losing its predictive value. In particular, the epochs at which a stock level rises to the value $S - s$ have a special role in the sense that they form a renewal process[*].

To determine $p_n(.)$, we introduce the two auxiliary functions $q_n(.)$ and $\omega_{n,m}(.)$.

---

[*] All these statements would be true even if the demands were to form a renewal process.

144

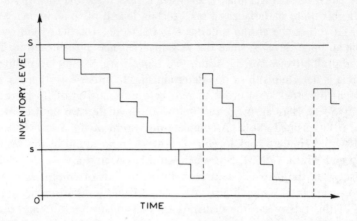

Fig. 8.1 The inventory level at a warehouse plotted against time

$$(8.3.7) \quad q_n(t) = \lim_{\Delta \to 0} \Pr\{X(t) = n \,|\, X(-\Delta) > X(0) = s\}$$

$$(8.3.8) \quad \omega_{n,m}(t) = \lim_{\Delta \to 0} \Pr\{X(t) = n \,|\, X(0) > X(-\Delta) = m\}.$$

By using elementary arguments, we find that

$$(8.3.9) \quad p_n(t) = \epsilon_{n,s}\, e^{-\lambda t}\, \frac{(\lambda t)^{S-n}}{(S-n)!} + \lambda \int_0^t e^{-\lambda u}\, \frac{(\lambda u)^{S-s-1}}{(S-s-1)!}\, q_n(t-u)\, du$$

$$(8.3.10) \quad q_n(t) = \int_t^\infty \alpha(v)\, dv\, e^{-\lambda t}\, \frac{(\lambda t)^{s-n}}{(s-n)!}\, (1 - \epsilon_{n,s})(1 - \delta_{n0})$$

$$+ \lambda \int_0^t \alpha(u)\, du \int_0^u e^{-\lambda v}\, \frac{(\lambda v)^{s-1}}{(s-1)!}\, dv\, \omega_{n,0}(t-u)$$

$$+ \sum_{m=0}^{s-1} \int_0^t \alpha(u)\, e^{-\lambda u}\, \frac{(\lambda u)^m}{m!}\, \omega_{n,s-m}(t-u)\, du$$

$$+ \delta_{n0}\, \lambda \int_0^t e^{-\lambda u}\, \frac{(\lambda u)^{s-1}}{(s-1)!}\, du \int_t^\infty \alpha(v)\, dv$$

$$(8.3.11) \quad \omega_{n,m}(t) = \epsilon_{n,s}\, e^{-\lambda t}\, \frac{(\lambda t)^{S-s+m-n}}{(S-s+m-n)!}\, (1 - \epsilon_{n,S-s+m})$$

$$+ \lambda \int_0^t e^{-\lambda u}\, \frac{(\lambda u)^{S-2s+m-1}}{(S-2s+m-1)!}\, q_n(t-u)\, du,$$

where $\epsilon_{n,s}$ is defined as

(8.3.12)
$$\epsilon_{n,s} = 1 \quad \text{for } n > s$$
$$= 0 \quad\quad n \leqslant s.$$

We note that any point t such that $X(t) = S$ is a regenerative event, and in view of this we have

(8.3.13)
$$\omega_{n,s}(t) = p_n(t),$$

a relation which can also be checked directly by putting $m = s$ in (8.3.11). The set of equations (8.3.9) through (8.3.11) can be solved by Laplace transform technique. After some calculations we obtain

(8.3.14) $q_n^*(\xi)$

$$= \left[(1 - \delta_{n0})(1 - \epsilon_{n,s}) \frac{\left(-\lambda \dfrac{\partial}{\partial \xi}\right)^{s-n}}{(s-n)!} \cdot \frac{1 - \alpha^*(\xi + \lambda)}{\xi + \lambda} + \delta_{n0}\rho_m^*(\xi + \lambda) \right.$$

$$+ \epsilon_{n,s}\, \alpha_m^*(\xi + \lambda) \frac{1}{\lambda} \left(\frac{\lambda}{\lambda + \xi}\right)^{S-s-n+1}$$

$$\left. + \epsilon_{n,s} \frac{1}{\lambda} \sum_{m=0}^{s-1} \frac{1}{m!} \left(\frac{\lambda}{\lambda + \xi}\right)^{S-s+m-n+1} \left(-\lambda \frac{\partial}{\partial \xi}\right)^m \alpha^*(\xi + \lambda) \right]$$

$$\times \left[1 - \left(\frac{\lambda}{\lambda + \xi}\right)^{S-2s} \alpha_m^*(\xi + \lambda) - \sum_{m=0}^{s-1} \left\{ \frac{1}{m!} \left(\frac{\lambda}{\lambda + \xi}\right)^{S-2s+m} \right.\right.$$

$$\left.\left. - \left(-\lambda \frac{\partial}{\partial \xi}\right)^m \alpha^*(\xi + \lambda) \right\} \right]^{-1},$$

where

(8.3.15) $\quad \alpha^*(\xi) = \int\limits_0^\infty e^{-\xi u}\, \alpha(u)\, du$

(8.3.16) $\quad \alpha_m^*(\xi) = \int\limits_0^\infty e^{-\xi u}\, \alpha(u)\, du \int\limits_0^u e^{-\lambda v} \frac{(\lambda v)^{m-1}}{(m-1)!} \lambda\, dv$

(8.3.17) $\begin{cases} \rho^*(\xi) = \dfrac{1 - \alpha^*(\xi)}{\xi}, \\[2mm] \rho_m^*(\xi) = \int\limits_0^\infty e^{-\xi u}\, \rho(u)\, du \int\limits_0^u e^{-\lambda v} \dfrac{(\lambda v)^{m-1}}{(m-1)!} \lambda\, dv. \end{cases}$

For any given form of $\alpha(.)$, $p_n(t)$ can be obtained by numerical inversion. With the help of $p_n(.)$, it is easy to carry out a cost analysis of the inventory (see problem 8.5, below).

8.4 Theory of dams

The discussion presented so far relates to the study of inventories of items measurable in discrete units (or lots). However, there is a class of problems in which the level of the inventory can be represented by a continuous variable. The simplest case is that of a dam characterized by the height. The inputs into the dam at different epochs, measured in terms of dam height, are also continuous random variables. Moran (1954, 1959) investigated some models of dams, and the theory was later developed by Prabhu (1964, 1965). A more recent account of the theory of dams can be found in the work of Cohen (1969). In this section we shall outline a general method of approach running on lines more or less parallel to the theory of queues. As observed in section **8.1**, the emphasis will be on the unification of methods that can be achieved by identifying imbedded regenerative processes, which in turn enables us to study all the significant features of the problem.

In a number of models that have been proposed so far, it is generally assumed that the epochs of input into the dam form a renewal process, the quantity of input at different epochs being continuously, independently, and identically distributed with common density function $\phi(.)$. Output (outflow) from the dam is assumed to be at unit rate[*] as long as the water-level is above zero, and output remains at zero when the dam level is zero. Output is resumed at unit rate as soon as the dam receives its next input. If the height of the dam is infinite, so that there is no overflow at any time, then a typical plot of the random variable $W(.)$ representing the water-level against t (see Fig. 6.3) is identical with that of $W(.)$ of queuing theory, discussed in section **8.2**. Thus the description of queues given in equation (8.2.6) can be taken *in toto* as the corresponding result for the theory of dams, provided we interpret the *service times* of individual customers as the *quantity of input*. The idle period in the queue, corresponding to the time-interval in which $W(t) = 0$, will now correspond to the dry period of the dam during which there is no output (outflow). The correlational structure of the bivariate point process comprising the epochs of commencement and termination of the dry periods can be readily deduced from the results of the previous section.

We next proceed to the case when the height of the dam is finite, equal to k. In such a case, when an input is such that the water-level exceeds k, there is an overflow of water that is in excess over the height k. In such a case $\pi(x, y, t)$ defined by (8.2.6) satisfies the backward integral equation

$$(8.4.1) \quad \pi(x, y, t) = H(y - t)\, \delta(x + t - y) \int_t^\infty f(u)\, du$$

$$+ \int_0^{\min(y,t)} f(u)\, du \int_0^\infty \phi(v)\, \pi(x, \min(k, y - u + v), t - u)\, dv$$

$$+ H(t - y) \int_y^t f(u)\, du \int_0^\infty \phi(v)\, \pi(x, \min(k, v), t - u)\, dv.$$

---

[*] There are other models in which the rate of outflow at any instant is a function of the water-level at that instant.

The above equation may appear somewhat difficult, but the Laplace transform solution can be obtained when the Laplace transforms of $f(.)$ and $\phi(.)$ are rational functions vanishing at ∞ faster than the reciprocal of their corresponding arguments. For full details the reader is referred to Srinivasan (1973).

8.5 Reliability theory

Reliability theory deals with general methods of evaluating the quality of systems that are subject to gradual deterioration or failure. It investigates the frequency of occurrence of defects in devices and methods of prediction. One general method of improving the reliability of a complex system consists in providing standby units. Even if a system is provided with a number of the unit that has deteriorated or failed, it cannot be guaranteed that the system does not break down due to failure of components. It may not be economically feasible to provide a sufficiently large number of standby units. Very often the system is provided with a few standby units, with facilities to repair the components that fail. Even then, immunity from system breakdown cannot be guaranteed, for there is always the possibility that the first unit that fails may not be repaired until the last available standby unit fails. Thus, from a practical point of view it is useful to study the characteristics of the system failures and also the time-intervals during which the system is operating efficiently, resulting in uninterrupted production. The time-interval in which the system is in failed condition, resulting in the cessation of all activities, is called the *down-time*. It is important to determine the statistical features of the cumulative down-time over a certain period, since it will help in making certain general policy decisions regarding the expansion of repair facilities and/or the provision of more standby units. A lucid account of general reliability theory can be found in the work of Gnedenko, Belyayev and Solovyev (1969). In this section, we shall illustrate the use of point processes in dealing with general problems of the reliability theory of repairable units. Reliability theory has a close connection with the theory of queues, to which we will allude presently. In fact, some of the problems of reliability theory have more general features and encompass some of the problems of queuing theory.

Let us consider an equipment which is being used for a production process. The equipment, due to the complex nature of the components of which it is made up, and whose characteristics in terms of quality and performance criterion cannot be predicted in a deterministic manner, fails after some time, its life-span of satisfactory performance being governed by the probability density function $f(.)$. We shall assume that n identical units are kept on standby, so that any one of them can be switched on to continue the production process. The switching on of a unit from the standby collection is assumed to be instantaneous, synchronizing with the event of failure of the unit engaged in the process of production. The units kept on standby are assumed to be failure-free. Such units are technically known as *cold standby units,* as contrasted with *warm* or *hot* standby units which can fail even in the state of being on standby. A repair facility is kept on alert so that it can attend to the repair of any unit that fails. The repair facility is such that it can attend to the repair of any one unit at a time, so that an item that is switched on from the standby collection and fails before the repair of the unit it replaced is completed, joins the queue at the repair counter. Thus we have a queue at this counter, the arrival pattern forming a renewal process conditional upon the system continuing the process of production. However, if there is a system breakdown, there is no further arrival to the queue. The inter-arrival times are distributed according to the probability density function $f(.)$. The renewal process of arrivals has a temporary pause with a system breakdown, and the process continues after the system re-starts the production process. If we assume that the repair times of different units are identically and independently distributed with common density function $g(.)$, essentially we have a queuing process of finite population, the departures being converted into arrivals after a random delay. Thus all the techniques of queuing theory and the imbedded regenerative processes are applicable in the present context.

For purposes of illustration let us first consider the case in which $g(.)$ is exponential with parameter λ. The epoch t is regenerative for the stochastic process $N(.)$, representing the number of items on standby, if t synchronizes with an item on standby just being switched on. We introduce the probability function $\pi_m(t)$ by

$$(8.5.1) \quad \pi_m(t) = \lim_{\Delta \to 0} \Pr\{Z(t) = 0 \,|\, N(0) > N(\Delta) = m\} \quad (m \leqslant n),$$

where $Z(t)$ is a random variable taking values 0 and 1 according to whether the system is or is not in a failed state. Concentrating our attention on the first regenerative event counted from the origin, we find that

$$(8.5.2) \quad \pi_m(t)$$

$$= \sum_{k=0}^{n-m-1} (m+k) \int_0^t f(u)\, e^{-\lambda u} \frac{(\lambda u)^k}{k!} \, \pi_{m+k-1}(t-u)\, du$$

$$+ n \int_0^t f(u)\, \pi_{n-1}(t-u)\, du \int_0^u e^{-\lambda v} \frac{(\lambda v)^{n-m-1}}{(n-m-1)!} \, \lambda\, dv \quad (0 < m < n)$$

$$(8.5.3) \quad \pi_0(t)$$

$$= e^{-\lambda t} \int_0^t f(u)\, du + \int_0^t f(u)\, du \int_u^t e^{-\lambda v} \lambda \pi_0(t-v)\, dv$$

$$+ \sum_{k=1}^{n-1} \int_0^t f(u) \, e^{-\lambda u} \frac{(\lambda u)^k}{k!} \, \pi_{k-1}(t-u) \, du$$

$$+ \int_0^t f(u) \, \pi_{n-1}(t-u) \, du \int_0^u e^{-\lambda v} \frac{(\lambda v)^{n-1}}{(n-1)!} \, \lambda \, dv.$$

Initially (at $t = 0$) one item is switched on and n items are on standby. Thus the principal quantity of interest is the function $\pi_n(t)$ which is now given by

$$(8.5.4) \qquad \pi_n(t) = n \int_0^t f(u) \, du \, \pi_{n-1}(t-u).$$

The Laplace transform solution of (8.5.3) and (8.5.2) can be readily obtained for any given n. Such a solution, combined with (8.5.4), yields explicitly the Laplace transform of $\pi_n(.)$. For instance, the mean value of $U(T)$ (known as "up time") during which the system is in the process of useful production is given by

$$E[U(T)] = E\left[\int_0^T Z(t) \, dt\right]$$

$$= \int_0^T E[Z(t)] \, dt$$

$$(8.5.5) \qquad = \int_0^T [1 - \pi_n(t)] \, dt.$$

To discuss the variance of $U(T)$, we need the function $\pi_n(t_1, t_2)$, where

$$(8.5.6) \quad \pi_n(t_1, t_2) = \lim_{\Delta \to 0} \Pr\{Z(t_1) = 0, Z(t_2) = 0 \mid N(\Delta) = n < N(0)\}.$$

For $t_1 < t_2$, by using the exponential nature of the repair-time distribution, we have

$$(8.5.7) \quad \pi_n(t_1, t_2) = \pi_n(t_1) \int_{t_1}^{t_2} e^{-\lambda v} \pi(t_2 - v) \, dv.$$

The variance of $U(T)$ is given by

$$(8.5.8) \quad \text{var}[U(T)] = 2 \int_0^T dt_2 \int_0^{t_2} \pi_n(t_1, t_2) \, dt_1 - \left[\int_0^T \pi_n(t) \, dt\right]^2.$$

Higher-order moments of $U(T)$ can be readily calculated (see Srinivasan and Gopalan (1974); see also problem 8.7, below).

Very often it is of interest to find out the time to the first system failure. Introducing the function $\pi_m^F(t)$ by

$$\pi_m^F(t)$$

$$= \lim_{\substack{\Delta \to 0 \\ \Delta' \to 0}} \Pr\{Z(t+\Delta) = 0, Z(u) \neq 0 \; \forall \, u \in [0, t] \mid N(\Delta') < N(0) = m\}/\Delta,$$

we note that $\pi_m^F(t)$ satisfies the equations

$$(8.5.9) \quad \pi_m^F(t) = \sum_{k=0}^{n-m-1} m \int_0^t f(u)\, e^{-\lambda u}\, \frac{(\lambda u)^k}{k!}\, \pi_{m+k-1}^F(t-u)\, du$$

$$+ m \int_0^t f(u)\, du \int_0^u e^{-\lambda v}\, \frac{(\lambda v)^{n-m-1}}{(n-m-1)!}\, \pi_{n-1}^F(t-u)\, dv, \quad 0 < m$$

$$(8.5.10) \quad \pi_0^F(t) = e^{-\lambda t} f(t) + \sum_{k=1}^{n-1} \int_0^t f(u)\, e^{-\lambda u}\, \frac{(\lambda u)^k}{k!}\, \pi_{k-1}^F(t-u)\, du$$

$$+ \int_0^t f(u)\, du \int_0^u e^{-\lambda v}\, \frac{(\lambda v)^{n-1}}{(n-1)!}\, \lambda\, dv\, \pi_{n-1}^F(t-u).$$

The principal quantity of interest is $\pi_n^F(t)$, known as the *reliability function*, which satisfies the relation

$$(8.5.11) \qquad \pi_n^F(t) = n \int_0^t f(u)\, \pi_{n-1}^F(t-u)\, du.$$

The Laplace transform solution of the above set of equations can readily be obtained for any given n.

In the analysis presented above we have assumed $g(.)$ to be an exponenti function of its argument. A similar analysis can be carried through for any $g(.)$ if $f(.)$ is an exponential function of its argument (see Srinivasan and Gopalan (1974)). It is desirable to extend the above type of analysis to the most general case when neither $f(.)$ nor $g(.)$ is of the exponential type. The stochastic process $Z(.)$ in such a case is non-Markovian and does not appear to have imbedded in it regenerative or renewal processes. While the queuing process discussed in section **8.3** does possess imbedded regenerative processes reliability analysis leads to a slightly more complicated queuing process at the repair counter, in which the arrivals are not independent. Viewed as a queuing process at the repair counter, the output is fed back into the repai counter after a certain random time-lag. It is this aspect of the problem that renders the solution difficult for $n > 2$. The case $n = 2$ is fairly simple and has been discussed by Srinivasan and Gopalan (1974).

There is yet another class of problems that remains relatively open. This corresponds to the case when the number of repair facilities exceeds one. In this case we have a many-channel queue with input and output mutually correlated. Only some simple models have been investigated so far (see, for example, Gnedenko *et al.* (1969), Kumagai (1971)).

CHAPTER 8: PROBLEMS

8.1 Show that the probability generating function $G(z, t)$ of the probabilities $\pi_n(t)$, defined by

$$G(z, t) = \sum_{n=0}^{\infty} \pi_n(t) z^n,$$

is given by

$$G^*(z, s) = \frac{z^{i+1} - (1-z)\xi^{i+1}/(1-\xi)}{(\lambda + \mu + s)z - \mu - \lambda z^2},$$

where $G^*(z, s)$ is the Laplace transform of $G(z, t)$ and ξ is the root with minimum modulus of the equation $\lambda z^2 + \mu - z(\lambda + \mu + s) = 0$. (Prabhu, 1965)

8.2 Show that the function $\pi_i(x, y, t)$ defined by

$$\pi_i(x, y, t) \, dx \, dy = \Pr \{L(t) = i, \ x < A(t) < x + dx, \ y < D(t) < y + dy\}$$

satisfies the equation

$$\frac{\partial \pi_i(x, y, t)}{\partial t} + \frac{\partial \pi_i(x, y, t)}{\partial x} + \frac{\partial \pi_i(x, y, t)}{\partial y}$$

$$= -\left(\frac{f(x)}{1 - F(x)} + \frac{\phi(y)}{1 - \Phi(y)}\right)\pi_i(x, y, t)$$

$$+ \delta(x) \int_{x'} \frac{f(x')}{1 - F(x')} \pi_{i-1}(x', y, t) \, dx' + \delta(y) \int_{y'} \frac{\phi(y')}{1 - \Phi(y')} \pi_{i+1}(x, y', t) \, dy',$$

where $F(.)$ and $\Phi(.)$ are the distribution functions of the inter-arrival and service times respectively.

8.3 Show that the above equation reduces to (8.2.4) when $f(.)$ and $\phi(.)$ are exponentially distributed with parameters λ and μ respectively.

8.4 Show that the function $p(x, t)$ defined by

$$p(x, t) = \Pr \{L(t) = n | \text{busy cycle started at } t = 0\}$$

is given by

$$p(x, t) = \frac{1}{\mu} \int_0^{\infty} \phi(y) \, dy \int_0^{\infty} \pi(x, y, t)\{\Phi_{n-1}(x) - \Phi_n(x)\} \, dx,$$

where $\mu = \int_0^{\infty} y \, \phi(y) \, dy$ and $\Phi_n(x)$ is the n-fold convolution of $\Phi(x)$, the distribution function of service times.

8.5 Show that the expected value of the sum of the inventory carrying cost and the shortage cost over a period t is given by

$$a \sum_{n=1}^{\infty} \int_0^t n \, p_n(u) \, du + b \int_0^t p_0(u) \, du,$$

where a is the inventory cost per item per unit time and b the shortage cost per item.

8.6 Consider a dam with finite capacity (say k units) in which the water-level $X(t)$ increases linearly at unit rate for a random length of time, after wh̵ it decreases linearly again at unit rate, the process alternating between period͏ of increase and decrease. The periods of increase and decrease are assumed to be exponentially distributed with parameters μ and λ respectively. If the function $\pi_0(x, y, t)$ is defined by

$$\pi_0(x, y, t) \, dx$$

$$= \Pr \{ x < X(t) < x + dx, X(u) > 0 \, \forall \, u \in [0, t] \, | \, X(0) = y, X(0^+) > y, X(0^-) >$$

show that the Laplace transform $\pi_0^*(x, s, p)$ of $\pi_0(x, y, t)$ is given by

$$\begin{vmatrix} \dfrac{a_1(s, p)}{D} & \dfrac{a_2(s, p)}{D} & \dfrac{a_3(x, s, p)}{D} - \pi_0^*(x, s, p) \\[2mm] a_1(s_0, p) & a_2(s_0, p) & a_3(x, s_0, p) \\[2mm] a_1(s_1, p) & a_2(s_1, p) & a_3(x, s_1, p) \end{vmatrix} = 0,$$

where

$$a_1(s, p) = \lambda \, e^{-(\lambda + p + s)k} (\lambda + p + s)(1 - e^{-(\mu + p + s)k}) - \lambda\mu \, e^{-(\lambda + p + s)k},$$

$$a_2(s, p) = -\mu(\lambda + p + s),$$

$$a_3(x, s, p) = \frac{\delta(x - k)}{\mu + p}(\lambda + p + s)(e^{-sk} - e^{-(\mu + p)k})$$

$$+ (\lambda + p + s) e^{-(\mu + p)x} (e^{(\mu + p - s)k} - 1)$$

$$+ e^{-sk}(\lambda + p + s)(1 - e^{-(\mu + p - s)k})(e^{-(\lambda + p)(k - x)})$$

$$+ \mu e^{-sx}(1 - e^{-(\lambda + p + s)(k - x)}),$$

$D = (\lambda + p + s)(\mu + p - s) - \lambda\mu$ and s_0 and s_1 are the two roots of the equation $D = 0$. (Srinivasan and Ramani, 1971)

8.7 Consider the n-unit system defined in section **8.5**. If the failure times are exponentially distributed with parameter μ, obtain the Laplace transform solution for $\rho(t)$, where

$\rho(t) = \Pr \{ Z(t) = 1$, given that one unit was switched on at $t = 0$ with $(n - 1)$ units on cold standby$\}$.

CHAPTER 8: APPENDIX

Derivation of equation (8.2.13) from (8.2.12)

The Laplace transform of the first term on the right-hand side of (8.2.12) is given by

$$\int\limits_0^\infty e^{-py}\,dy \int\limits_0^\infty e^{-st}\,dt \int\limits_t^\infty f(u)\,du\,\delta(x - \max(0, y - t))$$

$$= \int\limits_0^\infty e^{-py}\,dy \int\limits_0^y e^{-st}\,dt \int\limits_t^\infty f(u)\,du\,\delta(x - y + t)$$

$$+ \int\limits_0^\infty e^{-py}\,dy \int\limits_y^\infty e^{-st}\,dt \int\limits_t^\infty f(u)\,du\,\delta(x)$$

which, on integration, yields

$$e^{-px}\frac{[1 - f^*(p + s)]}{p + s} + \frac{\delta(x)}{p}\left[\frac{1 - f^*(s)}{s} - \frac{1 - f^*(p + s)}{p + s}\right].$$

The Laplace transform of the second term on the right-hand side of (8.2.12) is given by

$$\int\limits_0^\infty e^{-st}\,dt \int\limits_0^\infty e^{-py}\,dy \int\limits_0^\infty \phi(v)\,dv \int\limits_0^{\min(y, t)} f(u)\,\pi(x, y - u + v, t - u)\,du$$

$$= \int\limits_0^\infty e^{-st}\,dt \int\limits_0^\infty \phi(v)\,dv \int\limits_0^t e^{-py}\,dy \int\limits_0^y [\dots]\,du + \int\limits_t^\infty e^{-py}\,dy \int\limits_0^t [\dots]\,du.$$

By changing the order of integration, we find that the first term in the last expression equals

$$\int\limits_0^\infty \phi(v)\,dv \int\limits_0^\infty f(u)\,du \int\limits_u^\infty e^{-st}\,dt \int\limits_u^t \pi(x, y - u + v, t - u)\,e^{-py}\,dy,$$

which on change of variables becomes

$$\int\limits_0^\infty \phi(v)\,dv \int\limits_0^\infty f(u)\,du \int\limits_0^\infty e^{-s(t+u)}\,dt \int\limits_0^t \pi(x, y + v, t)\,e^{-p(y+u)}\,dy,$$

which, in turn, simplifies to

$$f^*(p + s) \int\limits_0^\infty \phi(v)\,dv \int\limits_y^\infty e^{-py}\,dy \int\limits_y^\infty \pi(x, y + v, t)\,e^{-st}\,dt.$$

Similarly, changing the order of integration in the second term, we find that it is equal to

$$\int\limits_0^\infty \phi(v)\,dv \int\limits_0^\infty f(u)\,du \int\limits_u^\infty e^{-py}\,dy \int\limits_u^y \pi(x, y - u + v, t - u)\,e^{-st}\,dt$$

which on change of variables becomes

$$\int_0^\infty \phi(v)\, dv \int_0^\infty f(u)\, du \int_0^\infty e^{-p(y+u)} \int_0^y e^{-s(t+u)}\, \pi(x, y+v, t)\, dt,$$

and this, in turn, reduces to

$$f^*(p+s) \int_0^\infty \phi(v)\, dv \int_0^\infty e^{-py}\, dy \int_0^y \pi(x, y+v, t)\, e^{-st}\, dt.$$

Thus the Laplace transform of the second term on the right-hand side of (8.2.12) is given by

$$f^*(p+s) \int_0^\infty \phi(v)\, dv \int_0^\infty e^{-py}\, dy \int_0^\infty \pi(x, y+v, t)\, e^{-st}\, dt$$

$$= \frac{f^*(p+s)}{2\pi i} \int_0^\infty \phi(v)\, dv \int_0^\infty e^{-py}\, dy \int_{\substack{\sigma-i\infty \\ \sigma>0}}^{\sigma+i\infty} \pi^*(x, p', s)\, e^{p'(y+v)}\, dp'$$

$$= \frac{f^*(p+s)}{2\pi i} \int_{\substack{\sigma-i\infty \\ 0<\sigma<\operatorname{Re} p}}^{\sigma+i\infty} \pi^*(x, p', s)\, dp' \int_0^\infty \frac{\phi(v)\, e^{p'v}\, dv}{p-p'}$$

$$= \frac{f^*(p+s)}{2\pi i} \int_{\substack{\sigma-i\infty \\ 0<\sigma<\operatorname{Re} p}}^{\sigma+i\infty} \frac{\pi^*(x, p', s)\, \phi^*(-p')\, dp'}{p-p'}.$$

The Laplace transform of the third term is given by

$$\int_0^\infty e^{-st}\, dt \int_0^t e^{-py}\, dy \int_0^t \phi(v)\, dv \int_0^t f(u)\pi(x, v, t-u)\, du,$$

which, on change of order of integration, becomes

$$\int_0^\infty \phi(v)\, dv \int_0^\infty f(u)\, du \int_u^\infty e^{-py}\, dy \int_u^\infty \pi(x, v, t-u)\, e^{-st}\, dt.$$

Proceeding as before, we find that the above expression equals

$$\int_0^\infty \phi(v)\, dv \int_0^\infty f(u)\, du \int_0^\infty e^{-py}\, dy \int_0^\infty \pi(x, v, t)\, e^{-s(t+u)}\, dt$$

$$= \int_0^\infty \phi(v)\, dv \int_0^\infty f(u)\, du\, e^{-su} \frac{(1-e^{-pu})}{p} \frac{1}{2\pi i} \int_{\substack{\sigma-i\infty \\ \sigma>0}}^{\sigma+i\infty} \pi^*(x, p', s)\, e^{p'v}\, dp'$$

$$= \frac{f^*(s)-f^*(p+s)}{p} \frac{1}{2\pi i} \int_{\substack{\sigma-i\infty \\ \sigma>0}}^{\sigma+i\infty} \pi^*(x, p', s)\phi^*(-p')\, dp'.$$

9   POINT PROCESSES IN BIOLOGY

9.1   Introduction

As has been observed in Chapter 2, population point processes were the subject of some of the earliest investigations made in the area of application of point processes. D.G. Kendall (1949), in a classic survey presented at a special Symposium of the Royal Statistical Society (London), outlined the method of dealing with stochastic fluctuations in the size of a general population. An account of such processes, with special reference to population point processes, can be found in the monograph of Harris (1963). Since then many significant contributions have been made towards the application of stochastic processes to biology, and it is not possible to include all of them in a single chapter of a monograph of this size. In this chapter, therefore, we discuss the application of point processes to problems that have relevance to current biological theories, particularly in the realms of cell structure and excitation phenomena. Even in such a restricted area there are many interesting papers, and we have preferred to deal only with those problems that serve to illustrate the theory and techniques developed in this monograph. The layout of this chapter is as follows. In section 9.2 we deal with the problem of burst-time distribution of a cell subject to viral attack. The next section contains an account of age-dependent population growth in relation to bacteriophage production. The concluding section gives an account of neuronal firings in terms of point processes and response phenomena.

9.2   Burst-time distribution of a cell

As we have observed in the introductory remarks, the use of stochastic methods in biology is of recent origin, particularly if we confine our attention to the basic problems of cell structure and their relation to bacteriophage production. A very good account of the mechanism of bacteriophage production and the stochastic models which explain the mechanism has been given by Gani (1965a). The mechanism of cell infection by the viral particle known as a *phage* is quite involved and follows a sequential pattern of several distinct stages. Let us assume that viral particles, say T-4 phages, are introduced into a suspension of *Escherichia coli* (bacteria). The bacterium, owing to its massive size compared with the phages, moves slowly in the suspension, while the T-4 particles (phages), due to Brownian motion, move rapidly until one or more of their tail fibres comes into contact with the wall of the bacterium. The phage gets itself anchored to the bacterium and the DNA strand of the phage is inserted into the cell. The attachment of a single phage to a bacterium marks the commencement of infection of the cell. Once the DNA strand has entered the bacterium, the normal reproductive cycle of the cell stops and, within the cell, acid and protein synthesis corresponding to the DNA strand of the viral particle commences. The period during which such a synthesis takes place is known as the *eclipse period*, and normally a considerable number of "vegetative phages", consisting of the DNA strand and some protein heads, tails and other parts that go to make up a viral particle, are produced. After the eclipse period, which lasts approximately 7−10 minutes, the vegetative phages get protein-coated. At the same time, more vegetative phages as well as protein coats are produced. The total DNA content of the cell is observed to vary linearly with the time elapsed since the virus particle got anchored to the cell. This process proceeds until the wall of the bacterium gives way, resulting in a burst, with the release of approximately 200−300 mature as well as 40−80 vegetative phages into the bacterial suspension. The culmination of the process is known as *lysis*. The mature phages repeat the parasitic cycle of growth in the manner described above, by first getting themselves attached to bacterial cells in the suspension. The vegetative phages contribute towards waste matter. In general, such a process results in the extinction of the bacterial colony in the suspension unless the cells, due to partial infection, produce phage-resistant mutants.

Gani and his co-workers (for an extensive bibliography see Gani (1965a)) developed a number of models which are based on these biological processes and describe the general behaviour of bacterial colonies subject to viral attack. Phage reproduction in a bacterium lends itself to formulation as a birth-and-death process. Since the replication of phages occurs only at the vegetative stage, the transition from the vegetative state to maturity can be conceived as death. The product-density approach is extremely useful in such a context, and this will be demonstrated in the next section. In this section we propose to deal with the distribution of time (measured from the epoch of infection) to lysis suffered by the bacterium.

We recall that once the DNA strand enters the bacterium, the growth of the bacterium and its reproductive capacity are arrested, and exact replicas of the DNA strand are produced within the cell. Thus the phage grows parasitically in the sense that it feeds on the cell material available in the bacterium. An empirical study of bacteria after phage attack suggests that the lethal effects depend on the number of phages living inside the bacterium and the total amount of toxic products produced by these phages since the attack. Thus it is reasonable to assume that lysis occurs when either the total number of phages inside the bacterium exceeds a certain number n_0, or the cumulative toxins produced exceed a certain quantity y_0. A model incorporating the dependence of the probability density function (governing the time to lysis) on the number of mature and vegetative phages was proposed by Puri (1969). The model takes into account only indirectly the effects due to toxic products. Another model which takes into account the effect due to toxins alone was studied by Srinivasan and Rangan (1970a). A reasonable measure of the cumulative toxic product in the bacterium can be represented by the random variable $Y(t)$, where

$$Y(t) = \int_0^t N(u)\,du$$

(9.2.1)
$$= \int_0^t (t-u)\,dN(u),$$

where $N(u)$ represents the number of vegetative phages in the interval $(0, u)$, so that the random variable $dN(u)$ can be defined by

(9.2.2) $dN(u) = \begin{cases} 1 & \text{if there is a phage birth in } (u, u+du), \\ 0 & \text{if there is no phage birth or phage} \\ & \text{maturation in } (u, u+du), \\ -1 & \text{if there is a phage maturation in} \\ & (u, u+du). \end{cases}$

We next observe that the random function $Y(t)$ is non-decreasing, and hence it is easy to restate our criterion for lysis in terms of probabilities. If T is the random variable representing the time to lysis, then $f(.)$, the probability density function of T, is given by

$$f(t) = \lim_{\Delta \to 0} \Pr\{N_1(t+\Delta) + N_2(t+\Delta) = n_0 > N_1(t) + N_2(t),\ Y(t) < y_0\}$$

(9.2.3)
$$+ \lim_{\Delta \to 0} \Pr\{Y(t) < y_0 < Y(t+\Delta),\ N_1(t) + N_2(t) < n_0\},$$

where $N_1(t)$ and $N_2(t)$ are the random variables representing the number of vegetative and mature phages present in the bacterium at time t. If the joint characteristic functional of the population processes $N_1(.)$ and $N_2(.)$ is known explicitly, $f(.)$ can be readily obtained. However, it is very difficult to derive the characteristic functional in an explicit form unless drastic (perhaps unrealistic) assumptions governing the growth of the phages are made (see, for example, Puri (1969), Srinivasan and Rangan (1970a)). For instance, if we assume that the birth and maturation probabilities are age-independent and equal to λ and μ respectively, then it is easy to deal with $\pi(n, m, y; t)$, where

(9.2.4) $\pi(n,m,y;t) = \lim_{\Delta \to 0} \Pr\{N_1(t) = n, N_2(t) = m, y < Y(t) < y + \Delta\},$

where the time origin synchronizes with the epoch of infection, which in turn is identified with the deposition of the DNA strand into the bacterium. For in this case, we can make use of the Markov nature of the process $\{N_1(.), N_2(.), Y(.)\}$ and obtain the forward differential equation

158

$$(9.2.5) \quad \frac{\partial \pi(n,m,y;t)}{\partial t} = -n(\lambda + \mu)\,\pi\,(n,m,y;t) + (n-1)\,\lambda\,\pi\,(n-1,m,y;t)$$

$$+ (n+1)\,\mu\,\pi\,(n+1,m-1,y;t) - (n-m)\,\frac{\partial \pi(n,m,y;t)}{\partial y},$$

with the initial condition

$$(9.2.6) \quad \begin{cases} \pi(n,m,y;0) \;=\; \delta(y) & n = 1,\, m = 0 \\ \qquad\qquad\quad = 0 & \text{otherwise.} \end{cases}$$

A further refinement is possible by identifying the time origin with the end of the eclipse period during which the maturation process is not possible. If the eclipse period is assumed to be of duration T_0, then (9.2.5) still holds good, provided we replace (9.2.6) by

$$\pi(n,m,y;0) \;=\; \omega(n,y,T_0) \qquad m = 0$$

$$\qquad\qquad\qquad = 0 \qquad\qquad \text{otherwise,}$$

where $\omega(n,y,T_0)$ is the joint probability frequency function of the process $\{N_1(.),\,Y(.)\}$ corresponding to a pure birth process with rate λ:

$$\omega(n,y,t) \;=\; \lim_{\Delta \to 0} \Pr\,\{N_1(t) = n,\, y < Y(t) < y + \Delta\}$$

which satisfies the equation

$$(9.2.7) \quad \frac{\partial \omega(n,y,t)}{\partial t}$$

$$= -n\lambda\omega(n,y,t) + (n-1)\lambda\omega\,(n-1,y,t) - n\,\frac{\partial\omega(n,y,t)}{\partial y}$$

$$(0 \leqslant t \leqslant T_0),$$

with the initial condition

$$(9.2.8) \quad \begin{cases} \omega(n,y,0) \;=\; \delta(y) & n = 1 \\ \qquad\qquad\quad = 0 & \text{otherwise.} \end{cases}$$

With the help of equations (9.2.6) through (9.2.8), it is easy to arrive at $f(.)$. If we are assuming $T_0 \neq 0$, then $f(t)$ will be identically equal to zero for $t \leqslant T_0$. Srinivasan and Rangan (1970a) dealt with (9.2.5) and (9.2.6) and obtained $f(.)$, as well as the expected number of mature phages that are produced due to lysis. Puri (1969) assumed $f(.)$ to be given by

$$f(t) = 0 \qquad\qquad\qquad 0 < t \leqslant T_0$$

$$= b(t)\,N_1(t) + c(t)\,N_2(t) \qquad t > T_0,$$

where $b(t)$ and $c(t)$ are arbitrary non-negative bounded functions, and he obtained estimates of the expected time to lysis and of the number of phages produced on lysis.

When the parameters of the birth-and-death process are either age- or time-dependent, the problem is indeed intractable if we are looking for an explicit solution. To account for such dependence, Gani and his co-workers (see Gani (1965a)) have used the fixed-size pool model. In this model it is assumed that the number of survivors n, having the capacity to reproduce (vegetative type), is a constant, a new vegetative phage directly replacing a mature phage on its departure from the pool. The replacements are supposed to occur at intervals that are independently and identically distributed. Such a model appears to bring to the fore the age-dependent features. For details the reader is referred to the survey article by Gani (1965a).

9.3 Age-dependent models for phage reproduction

In the theory of bacteriophage reproduction and its parasitic cycle of growth, one of the problems of interest is the distribution of the number of vegetative and mature phages released at the time of lysis. The previous section contains a method of estimating the moments of the burst size distribution, when the probabilities of birth and maturation are age independent. However, if the lysis time is assumed to be deterministic, it is possible to determine the moments of the burst size distribution. To do this we make the following assumptions:

(i) The replication of the DNA injected into the bacterium and the consequent formation of new phages is an age-dependent birth process during the period $(0, T_0)$, where T_0 is the eclipse period, the equivalent age of the phage at $t = 0$ being x_0.

(ii) The sub-populations generated by two co-existing phages develop in complete independence of one another.

(iii) A phage of age x at time t has a probability

$$R(x', x'' | x) \, dx' dx'' dt + o(dt)$$

of splitting in $(t, t + dt)$ into two vegetative phages of ages lying in the intervals $(x', x' + dx')$, $(x'', x'' + dx'')$.

Similarly, in (T_0, T) the given primary of age x at $t = T_0$ generates a population in accordance with assumptions (ii) and (iii), along with

(iv) A vegetative phage of age x at t has a probability $\mu(x) dt + o(dt)$ of becoming a mature phage in $(t, t + dt)$.

(v) The birth-and-death probabilities $R(x', x'' | x) \, dx' dx''$ and $\mu(x)$ depend only on the age x of the phage and not on t, the time of its existence.

We shall obtain the first two moments of the number of vegetative phages above a certain age x at any time t.

Let

$$f_v^{(1)}(x|x_0, T_0), \; g_v^{(1)}(x|x_0, T-T_0)$$

and

$$f_v^{(2)}(x_1, x_2|x_0, T_0), \; g_v^{(2)}(x_1, x_2|x_0, T-T_0)$$

be the product densities of degrees one and two of vegetative phages in the intervals $(0, T_0)$ and (T_0, T).

$f_v^{(1)}(x|x_0, T_0)dx$ denotes the probability that there exists one vegetative phage of age between x and $x + dx$ at time $t = T_0$, due to a primary of age x_0 at $t = 0$, and $g_v^{(1)}(x|x_0, T-T_0)$ denotes the probability that there exists one vegetative phage of age between x and $x + dx$ at T, due to a primary of age x_0 at $t = T_0$. Likewise, $f_v^{(2)}(x_1, x_2|x_0, T-T_0)dx_1dx_2$ denotes the probability of observing one vegetative phage having an age between x_1 and $x_1 + dx_1$ and another having an age between x_2 and $x_2 + dx_2$ at $T-T_0$, starting with a vegetative phage of age x_0 at $t = 0$, while $g_v^{(2)}(x_1, x_2|x_0, T-T_0)$ is the probability of observing the same thing at $t = T$, starting with a primary at $t = T_0$.

To obtain the equations for the product densities in $(0, T_0)$, that is during the eclipse period, we define the characteristic functional for the process as

$$(9.3.1) \quad C(\theta(x), T_0|x_0) \; = \; \mathrm{E}\left[\exp\left\{\int_x \theta(x)d_x N(x; T_0)\right\}\right].$$

The product densities are simply obtained as the functional derivatives of the characteristic functional. Thus we have

$$(9.3.2) \quad (-i)^h \left.\frac{\partial^h C(\theta(x), T_0|x_0)}{\partial\theta(x_1)\,\partial\theta(x_2)\ldots\partial\theta(x_h)}\right|_{\theta=0} = f_h(x_1 \ldots x_h),$$

where $f_h(x_1 \ldots x_h)$ is the product density of order h.

Using the regenerative property of the process, we can immediately write down the renewal-type equation for the characteristic functional (see Bellman and Harris (1948)) as

$$(9.3.3) \quad C(\theta(x), T_0|x_0)$$

$$= \exp\{i\theta(x_0 + T_0)\}\exp\left\{-\int_0^{T_0}[\,\iint R(x', x''|x_0 + u)dx'dx'']du\right\}$$

$$+ \int_0^{T_0} [\iint R(x', x''|x_0 + u)C(\theta(x), T_0-u|x')C(\theta(x), T_0-u|x'')$$

$$\times \exp\left\{-\int_0^u [\iint R(y', y''|x_0 + v)dy'dy'']\,dv\right\}dx'dx''\,du,$$

where the first term on the right-hand side takes into account the case where no event takes place in $(0, T_0)$, and the second the splitting of a vegetative phage of age x into two of ages x' and x''. Using (9.3.2)

$$(9.3.4) \quad f_v^{(1)}(x|x_0, T_0) \exp \int_0^{T_0} [\iint R(x', x''|x_0 + T_0 - \tau')dx'dx''] \, d\tau'$$

$$= \int_0^{T_0} [\iint R(x', x''|x_0 + T_0 - \tau')dx'dx''] [f_v^{(1)}(x|x', \tau') + f_v^{(1)}(x|x'', \tau')]$$

$$\times \exp \{\int_0^{\tau'} [\iint R(x', x''|x_0 + T_0 - \tau'')dx'dx'']d\tau''\}d\tau' + \delta(x - x_0 - T_0)$$

and

$$(9.3.5) \quad f_v^{(2)}(x_1, x_2|x_0, T_0) \exp - \int_0^{T_0} [\iint R(x', x''|x_0 + T_0 - \tau')dx'dx'']d\tau'$$

$$= \int_0^{T_0} [\iint R(x', x''|x_0 + T_0 - \tau')dx'dx''] [f_v^{(2)}(x_1, x_2|x', \tau')$$

$$+ f_v^{(2)}(x_1, x_2|x'', \tau') + f_v^{(1)}(x_1|x', \tau')f_v^{(1)}(x_2|x'', \tau')$$

$$+ f_v^{(1)}(x_1|x'', \tau')f_v^{(1)}(x_2|x', \tau')]$$

$$\times \{\exp - \int_0^{\tau'} [\iint R(x'x''|x_0 + T_0 - \tau'')dx'dx'']d\tau''\}d\tau'.$$

For the evolution of the process in (T_0, T), the initial condition will be a spectrum to be determined by $f_v^{(1)}(x|x_0, T_0), f_v^{(2)}(x_1, x_2|x_0, T_0) \ldots$. Since the spectrum is specified by the product densities of the first few orders, it will not in itself be sufficient to determine the characteristic functional. But this is not a serious drawback if we confine ourselves only to the determination of the first few moments.

Now denoting by $F_v^{(1)}(x|x_0, T)$ the probability that there exists one vegetative phage of age between x and $x + dx$ at $t = T$ due to a primary of age x_0 at $t = 0$,

$$(9.3.6) \quad F_v^{(1)}(x|x_0, T) = \int_{x_1} f_v^{(1)}(x_1|x_0, T_0)g_v^{(1)}(x|x_1, T - T_0)dx_1,$$

where the integration is taken over all values of x_1. Here we have made use of the homogeneous nature of the process in (T_0, T) with respect to time, so that $g_v^{(1)}(x|x_0, T - T_0)$ is a function of $T - T_0$ only.

Similarly, if $F_v^{(2)}(x_1, x_2|x_0, T)$ denotes the probability that there exist two vegetative phages, one of age x_1 and the other of age x_2 at $t = T$ due to a primary of age x_0 at $t = 0$, then

$$(9.3.7) \quad F_v^{(2)}(x_1, x_2|x_0, T) = \int_{x'} f_v^{(1)}(x'|x_0, T_0)g_v^{(2)}(x_1, x_2|x', T - T_0)dx'$$

$$+ \int_{x'} \int_{x''} f_v^{(2)}(x', x''|x_0, T_0)g_v^{(1)}(x_1|x', T - T_0)$$

$$\times g_v^{(1)}(x_2|x'', T - T_0)dx'dx''.$$

The first term on the right-hand side is due to the fact that the two phages of ages x_1 and x_2 at T originate from a primary of age x' at $t = T_0$, which itself has an ancestor of age x_0 at $t = 0$. The second term arises from the fact that at $t = T_0$ we can find two individuals of ages x' and x'' due to a primary of age x_0 at $t = T_0$, and each of these in turn gives rise to phages of age x_1 and x_2 at $t = T$.

Proceeding in the same way as in the determination of $f_V^{(1)}(x|x_0, T_0)$ and $f_V^{(2)}(x_1, x_2|x_0, T_0)$, we write the equations governing $g_V^{(1)}$ and $g_V^{(2)}$:

$$g_V^{(1)}(x|x_0, T - T_0) \exp \int_{T_0}^{T} [\iint R(x', x''|x_0 + T - T') dx' dx'' + \mu(x_0 + T - T')] dT'$$

$$= \int_{T_0}^{T} [\iint R(x', x''|x_0 + T - T') dx' dx''] [g_V^{(1)}(x|x', T' - T_0) + g_V^{(1)}(x|x'', T' - T_0)]$$

$$\times \exp \{\int_{T_0}^{T'} [\iint R(x', x''|x_0 + T - T'') dx' dx'' + \mu(x_0 + T - T'')] dT''\} dT'$$

$$(9.3.8) \quad + \delta(x - x_0 - T + T_0),$$

$$(9.3.9) \quad g_V^{(2)}(x_1, x_2|x_0, T - T_0)$$

$$\times \exp \int_{T_0}^{T} [\iint R(x', x''|x_0 + T - T') dx' dx'' + \mu(x_0 + T - T')] dT'$$

$$= \int_{T_0}^{T} [\iint R(x', x''|x_0 + T - T') dx' dx'' + \mu(x_0 + T - T')]$$

$$\times [g_V^{(2)}(x_1, x_2|x', T' - T_0) + g_V^{(2)}(x_1, x_2|x'', T' - T_0)$$

$$+ g_V^{(1)}(x_1|x', T' - T_0) g_V^{(1)}(x_2|x'', T' - T_0)$$

$$+ g_V^{(1)}(x_1|x'', T' - T_0) g_V^{(1)}(x_2|x', T' - T_0)]$$

$$\times \exp \{\int_{T_0}^{T'} [\iint R(x', x''|x_0 + T - T'') dx' dx'' + \mu(x_0 + T - T'')] dT''\} dT$$

Thus, once the product densities $g_V^{(1)}, g_V^{(2)}, \ldots, f_V^{(1)}, f_V^{(2)}, \ldots$, are determined, the product densities for the whole process in $(0, T)$, namely $F_V^{(1)}$, $F_V^{(2)}, \ldots$, can immediately be written down using (9.3.6) and (9.3.7). It is interesting to note that the integral equations for $f_V^{(1)}, f_V^{(2)}, g_V^{(1)}$ and $g_V^{(2)}$ are linear and hence may be solved for any given form of R. Explicit solutions for the various product densities have been obtained by Srinivasan and Rangan (1970b) for certain special forms of $R(x', x''|x_0)$.

So far, we have confined our attention to bacteria attacked by a single viral onset. It is possible, however, that two or more phages get attached to the wall of a bacterial cell. This aspect of the problem has attracted considerable attention (see Gani (1965b)). Following Gani, let us consider the adsorption process for phages after they are inserted into a bacterial suspension. At an initial time $t_0 < 0$ we start with one bacterium, so that at $t = 0$ the mean number of bacteria is N. Then x_0 phages ($x_0 < N$) are introduced into the bacterial suspension at $t = 0$. We assume (i) that at most s phages can get attached to any bacterium, and (ii) that for each collision between a bacterium and a phage, attachment will depend on a parameter $\lambda_i > 0$ ($i = 0, 1, \ldots, s$) that varies with the number i of phages already attached to the particular bacterium. Gani studied this model and dealt with the joint distribution of the number n_i of bacteria with i attachments and $x(t)$ the number of unattached phages. The results obtained by Gani will be extremely useful if we confine our attention to time-intervals that are not large compared with the eclipse period. However, if the time-interval is large enough to allow one or more lyses to materialize, then the number of unattached phages is no longer deterministic, since a random number of mature phages gets added to the suspension with each lysis. It is in this context that the kinetic equations introduced in Chapter 7 and suitably adapted for the present situation may be useful. The above assumptions will result in an open hierarchical system of equations for the product densities of bacteria with specified numbers of attachments and specified numbers of vegetative and mature phages within (see Srinivasan and Rangan (1970a) for details). Such an approach has the obvious advantage that all the aspects of the bacterium—phage interaction are included. Although the resulting equations may appear to be intractable, it may be advantageous to use some closure approximation similar to the one used in kinetic theory (see Chapter 7) and seek the first two moments of the distribution of the number of bacteria with specified characteristics. This will also lead to the determination of the probability of the time T to extinction of the bacterial colony.

9.4 Neurophysiology

The central nervous system of the higher animals consists of large collections of individual nerve cells or neurons, all strongly interacting with one another. The best way to study the behaviour of such systems is to resort to models depicting the characteristics of intricate nets of artificial elements whose behaviour is related to that of physiological neurons. The interval histograms of spontaneous-active lateral geniculate neurons obtained experimentally by Bishop, Levick and Williams (1964) show that a stochastic approach may be very helpful in understanding nervous response phenomena. Several models of point processes have been proposed. Ricciardi and Esposito (1966) characterized the responses as Poisson events suitably censored. Ten Hoopen and Reuver (1965) suggested a model based on the interaction of an inhibitory renewal process and an excitatory renewal process. A slightly different approach due to Gerstein and Mandelbrot (1964) consists in identifying the spike activity of the neuron with the random walk problem. This approach has the advantage that the model brings out the idea of neuronal post-synaptic potential in a natural way. In this section we shall briefly discuss point process models. Since a detailed analysis of interacting renewal processes with special reference to the ten Hoopen–Reuver model has already been presented in Chapter 6 in connection with bivariate processes, we shall in this section discuss other models.

Ten Hoopen and Reuver (1968) proposed a model of neuronal spike trains based on selective interaction between two dependent recurrent time-sequences of stimuli, called "excitatory" and "inhibitory". The excitatory process (hereinafter called the ϕ-*process*) is assumed to be a stationary renewal point process. Each ϕ-event triggers, in a manner to be specified later, an inhibitory point process called the ψ-*process*. The sequence of ψ-events originating from a particular ϕ-event continues until the next ϕ-event, the ϕ-event being deleted by this sequence of ψ-events. Every undeleted event is assumed to give a response (called a *p-event*), the central quantity of interest being the probability frequency function of two successive response-yielding events.

Ten Hoopen and Reuver (1968) actually proposed two models, spelling out the mechanism of interaction between the excitatory and inhibitory processes. In model I, it is assumed that every ϕ-event, deleted or not, triggers a sequence of ψ-events such that the interval between that ϕ-event and the next ψ-event, as well as the intervals between successive ψ-events, are independently and identically distributed with probability frequency function $\psi(t)$. In model II, it is assumed that only undeleted ϕ-events can give rise to inhibitory processes. Explicit solutions for the renewal densities of the p-event were obtained by Srinivasan and Rajamannar (1970a). We now briefly indicate the various steps leading to the renewal densities.

Model I of ten Hoopen and Reuver

Let $h_r(t)dt$ denote the probability that a p-event occurs in $(t, t + dt)$. Let us assume without loss of generality that a p-event occurs at $t = 0$. We wish to obtain $p(t)$, where $p(t)dt$ denotes the probability that the next response-yielding event occurs between t and $t + dt$. At the outset, we observe that the p-events constitute a renewal process, since the information prior to the occurrence of a p-event is irrelevant for the future evolution of the system. In such a case, it is sufficient if we determine the renewal density of p-events.

Let $h_r(t)$ denote the renewal density of p-events. The density $h_r(t)$ can be identified with the product density of degree one in a more general case, when the events do not constitute a renewal process. If $g(t)$ is the renewal density of ϕ-events, we observe that

(9.4.1) $h_r(t)dt$

$$= \phi(t)[1 - \int_0^t \psi(t')dt']\,dt + \int_0^t g(\tau)\phi(t-\tau)[1 - \int_0^{t-\tau} \psi(t')dt']\,dtd\tau,$$

where $\psi(t)dt$ denotes the probability that the first inhibitory event due to a ϕ-event at $t = 0$ occurs in $(t, t + dt)$.

The foregoing equation is obtained by arguing that the response at t is either the first excitatory event or a subsequent event.

Taking the Laplace transform with respect to t on both sides of (9.4.1), we have

$$(9.4.2) \qquad h_r^*(s) = K^*(s) + g^*(s)K^*(s),$$

where

$$(9.4.3) \qquad K(t) = \phi(t)[1 - \int_0^t \psi(t')dt'].$$

Now $g(t)$ can be expressed in terms of the interval distribution of the excitatory events (see Chapter 3):

$$(9.4.4) \qquad g(t) = \phi(t) + \int_0^t g(t')\phi(t-t')dt'.$$

The Laplace transform solution of (9.4.4) is given by

$$(9.4.5) \qquad g^*(s) = \frac{\phi^*(s)}{1 - \phi^*(s)}$$

Using (9.4.5), we find that

$$(9.4.6) \qquad p^*(s) = \frac{K^*(s)}{1 - \phi^*(s) + K^*(s)}.$$

For the special case when the excitatory events constitute a Poisson process with parameter μ, we obtain

$$(9.4.7) \qquad p^*(s) = \mu[1 - \psi^*(s+\mu)][s + \mu - \mu\psi^*(s+\mu)]^{-1},$$

a result obtained by ten Hoopen and Reuver. It should be noted, however, that this result is independent of the interval distribution of two successive events. This is as it should be, since the deletion of a ϕ-event depends only on the formation of the first inhibitory pulse and not on the subsequent pulses.

Model II of ten Hoopen and Reuver

In this case, it is very easy to obtain the interval distribution of p-events. We notice that two successive p-events can be intercepted by at most one ϕ-event. Making use of this fact and of the renewal nature of the process, we observe that

$$(9.4.8) \qquad p(t) = \phi(t)[1 - \int_0^t \psi(t')dt'] + \int_0^t \phi(\tau)\phi(t-\tau)\int_0^\tau \psi(t')dt'd\tau,$$

so that we have

$$(9.4.9) \qquad p^*(s) = K^*(s) + \phi^*(s)[\phi^*(s) - K^*(s)].$$

For the special case when the excitatory process constitutes a Poisson process, we have

$$(9.4.10) \quad p^*(s) = \frac{\mu s}{(\mu + s)^2} [1 - \psi^*(\mu + s)] + \left(\frac{\mu}{\mu + s}\right)^2.$$

Once we are in possession of the interval distribution of p-events, it is easy to obtain the renewal density of p-events by using relations similar to (9.4.4) and (9.4.5). With the help of the renewal density, it is possible to obtain all the moments of the number of p-events during a certain time-interval $(0, t)$.

An equation similar to (9.4.8) can be obtained even if the excitatory events constitute a stationary point process. In this case, the excitatory process is described by the sequence of interval distributions $\phi_1(t), \phi_2(t_1, t_2), \ldots$ where $\phi_2(t_1, t_2) dt_1 dt_2$ denotes the probability that two successive events happen, one in $(t_1, t_1 + dt_1)$ and the other in $(t_2, t_2 + dt_2)$, given that an even has happened at $t = 0$. It is very interesting to observe that $p(t)$, the interval distribution between any two successive p-events, is given by

$$(9.4.11) \quad p(t) = \phi_1(t)[1 - \int_0^t \psi(t')dt'] + \int_0^t \phi_2(t_1, t) \int_0^{t_1} \psi(\tau)d\tau \, dt_1,$$

where we have made use of the stationary nature of the process and $\psi(t)$ has the same interpretation as in (9.4.8). In a similar manner, higher-order interval distributions can be obtained. It should be noted, however, that in the case of model I such a simple treatment is not possible in view of the non-Markovian nature of the problem, arising from the triggering of inhibitory pulses by all ϕ-events.

It is not very realistic to imagine that an inhibitory event can annihilate the next ϕ-event irrespective of its time of arrival. In fact, ten Hoopen (1966) has pointed out that it might be useful to introduce a criterion for the effectiveness of the inhibitory pulses as a function of time. We will assume that the inhibitories are effective only for a time τ after the formation of the first pulse, τ being governed by the probability frequency function $\psi_2(\tau)$. Let $\psi_1(t)$ be the probability frequency function of the interval between any ϕ-event and the next ψ-event due to it. We notice that with this modification the undeleted events constitute a renewal point process in either of the model of ten Hoopen and Reuver. Thus we can confine our attention to the interval distribution of undeleted events or to the first-order product density of undeleted events. Srinivasan and Rajamannar (1970b) have studied this model and obtained the interval characteristics of the undeleted events. There is yet another aspect worth mentioning. In many cases, the individual neuron is too weak to be capable of being observed; in fact the undeleted events get stored, and as soon as the number of such stored events exceeds a threshold value

n, a response occurs. This aspect of the problem has been analysed by ten
Hoopen and Reuver (1967a) and Srinivasan, Rajamannar and Rangan (1971),
who have dealt with the interval and other higher-order characteristics of the
response-yielding events.

REFERENCES

Arrow, K.J., Karlin, S. and Scarf, H. (1958), *Studies in the Mathematical Theory of Inventory and Production*, Stanford University Press.

Bartlett, M.S. (1954), Processus stochastiques ponctuels, *Ann. Inst. H. Poincaré*, **14**, 35–60.

Bartlett, M.S. (1963), The spectral analysis of point processes, *J. Roy. Statist. Soc.*, **B, 25**, 264–96.

Bartlett, M.S. (1966), *An Introduction to Stochastic Processes* (2nd edn), Cambridge University Press.

Bartlett, M.S. and Kendall, D.G. (1951), On the use of the characteristic functional in the analysis of some stochastic processes in physics and biology, *Proc. Camb. Phil. Soc.*, **47**, 65–76.

Bedard, G. (1966), Photon counting statistics of Gaussian light, *Phys. Rev.*, **151**, 1038–9.

Bellman, R.E. and Harris, T.E. (1948), On the theory of age dependent stochastic branching processes, *Proc. Nat. Acad. Sci.* (U.S.A.), **34**, 601–4.

Beran, M. (1968), *Statistical Continuum Theories*, John Wiley & Sons, New York.

Beutler, F.J. and Leneman, O.A.Z. (1966), The theory of stationary point processes, *Acta Math.*, **116**, 159–97.

Bhabha, H.J. (1950), On the stochastic theory of continuous parametric systems and its application to electron–photon cascades, *Proc. Roy. Soc. Lond.*, **A, 202**, 301–32.

Bhabha, H.J. and Heitler, W. (1937), The passage of fast electrons and the theory of cosmic showers, *Proc. Roy. Soc. Lond.*, **A, 159**, 432–58.

Bharucha-Reid, A.T. (1960), *Elements of the Theory of Markov Processes and their Applications*, McGraw-Hill, New York.

Bishop, P.O., Levick, W.R. and Williams, W.O. (1964), Statistical analysis of the dark discharge of lateral geniculate neurons, *J. Physiol.* (London), **170**, 598–612.

Blackwell, D. (1948), A renewal theorem, *Duke Math. J.*, **15**, 145–50.

Blackwell, D. (1953), Extension of a renewal theorem, *Pacific J. Math.*, **3**, 315–20.

Blanc-Lapierre, A., Dumontet, P. and Picinbono, B. (1965), Study of some statistical models introduced by problems of physics, Bernoulli–Bayes–Laplace Anniversary Volume, Proc. Int. Res. Seminar, Berkeley, 1963, pp. 9–16.

Bochner, S. (1947), Stochastic processes, *Ann. Math.*, **48**, 1014–61.

Bochner, S. (1955), *Harmonic Analysis and the Theory of Probability*, University of California Press, Berkeley.

Bogoliubov, N.N. (1962), Problems of a Dynamical Theory in Statistical Physics, in De Boer, J. and Uhlenbeck, G.E. (editors), *Studies in Statistical Mechanics*, vol. 1, North-Holland Publishing Co., Amsterdam, pp. 5–116.

Chow, Y.S. and Robbins, H. (1963), A renewal theorem for random variables which are dependent or non-identically distributed, *Ann. Math. Statist.*, **34**, 390–5.

Cinlar, E. (1968), On the superposition of *m*-dimensional point processes, *J. Appld Prob.*, **5**, 169–76.

Cohen, E.G.D. (1962), The Boltzmann equation and its generalization to higher densities, in Cohen (ed.), *Fundamental Problems in Statistical Mechanics*, vol. 1, North-Holland Publishing Co., Amsterdam.

Cohen, E.G.D. (1968), Kinetic Theory of Dense Gases, in Cohen (ed.), *Fundamental Problems in Statistical Mechanics*, vol. 2, North-Holland Publishing Co., Amsterdam, pp. 228–75.

170

Cohen, J.W. (1969), *The Single Server Queue*, North-Holland Publishing Co., Amsterdam.
Cox, D.R. (1955), Analysis of non-Markovian stochastic processes by the inclusion of supplementary variables, *Proc. Camb. Phil. Soc.*, **51**, 433–41.
Cox, D.R. (1962), *Renewal Theory*, Methuen, London.
Cox, D.R. and Lewis, P.A.W. (1966), *The Statistical Analysis of Series of Events*, Methuen, London.
Cox, D.R. and Lewis, P.A.W. (1970), Multivariate point processes, *Proc. 6th Berkeley Symp. Math. Statist. and Prob.* (to be published).
Daley, D.J. (1968), The correlation structure of the output process of some single server queuing systems, *Ann. Math. Statist.*, **39**, 1007–19.
Doob, J.L. (1948), Renewal theory from the point of view of the theory of probability, *Trans. Amer. Math. Soc.*, **63**, 422–38.
Feller, W. (1941), On the integral equation of renewal theory, *Ann. Math. Statist.*, **12**, 243–67.
Feller, W. (1948), On probability problems in the theory of counters, *Courant Anniversary Volume*, Interscience Publishers, New York, 105–15.
Feller, W. (1949), Fluctuation theory of recurrent events, *Trans. Amer. Math. Soc.*, **67**, 98–119.
Feller, W. (1966), *An Introduction to Probability Theory and its Applications*, vol. 2, John Wiley & Sons, New York.
Feller, W. (1968), *An Introduction to Probability Theory and its Applications*, vol. 1, 3rd edn, John Wiley & Sons, New York.
Fry, T.C. (1928), *Probability and its Engineering Uses*, Van Nostrand, New York.
Furry, W.H. (1937), On fluctuation phenomena in the passage of high energy electrons through lead, *Phys. Rev.*, **52**, 569–81.
Gani, J. (1965a), Stochastic models for bacteriophage, *J. Appld Prob.*, **2**, 225–68.
Gani, J. (1965b), Stochastic phage attachment to bacteria, *Biometrics*, **21**, 134–9.
Gaver, D.P. (1963), Random hazard in reliability problems, *Technometrics*, **5**, 211–26.
Gerstein, C.D. and Mandelbrot, D. (1964), Random walk models for the spike activity of a single neuron, *Biophys. J.*, **4**, 41–68.
Gnedenko, B.V. (1958), *The Theory of Probability* (translation of 2nd edition), Chelsea Publishing Co., New York.
Gnedenko, B.V., Belyayev, Yu.K. and Solovyev, A.D. (1969), *Mathematical Methods of Reliability Theory* (translated edition), Academic Press, New York.
Green, H.S. (1952), *Molecular Theory of Fluids*, North-Holland Publishing Co., Amsterdam
Hadley, G. and Whitin, T.M. (1963), *Analysis of Inventory Systems*, Prentice Hall Inc., Englewood Cliffs, New Jersey.
Hammersley, J.M. (1953), On counters with random dead time: I, *Proc. Camb. Phil. Soc.*, **49**, 623–37.
Harris, T.E. (1963), *The Theory of Branching Processes*, Springer Verlag, Berlin.
Hawkes, A.G. (1971), Spectra of some self-exciting and mutually exciting point processes, *Biometrika*, **58**, 83–90.
Hille, E. (1962), *Analytic Function Theory*, vol. 2, Blaisdell Publishing Co., New York.
Hille, E. and Phillips, R.S. (1957), *Functional Analysis and Semi-Groups*, Amer. Math. Soc. Colloq. Publications, vol. 31.
Ichimaru, S. (1968), Theory of turbulent state of a plasma, *Phys. Rev.*, **165**, 231–50.
Ichimaru, S. (1970), Theory of Strong Turbulence in Plasmas, *Phys. of Fluids*, **13**, 1560–72.
Jakeman, E. and Pike, E.R. (1968), The intensity-fluctuation distribution of Gaussian light, *J. Phys.*, Ser. 2, **1 A**, 128–38.
Janossy, L. (1950), On the absorption of a nucleon cascade, *Proc. Roy. Irish Acad. Sci.*, **A 53**, 181–8.
Karp, S. and Clark, J.R. (1970), Photon counting: a problem in classical noise theory, *IEEE Trans. on Inf. Theory*, *IT–16*, 672–80.
Kemeny, J.G. and Snell, J.L. (1960), *Finite Markov Chains*, Van Nostrand, Princeton, N.J.

171

Kendall, D.G. (1949), Stochastic processes and population growth, *J. Roy. Statist. Soc.*, **B 11**, 230–64.
Kendall, D.G. (1951), Some problems in the theory of queues, *J. Roy. Statist. Soc.*, **B 13**, 151–85.
Kendall, D.G. (1964), Some recent work and further problems in the theory of queues, *Theory of Prob. and Applications* (trans.), **9**, 1–13.
Khintchine, A.Y. (1955), *Matematicheskie metody teorii massovogo obsluzhivania*, Trudy Matematicheskovo Instituta Steklov, Akad. Nauk, U.S.S.R., **49**. Translated by D.M. Andrews and M.H. Quenouille under the title *Mathematical Methods in the Theory of Queueing*, Griffin, London, 1960 (2nd edn, 1969).
Khintchine, A.Y. (1960), *Mathematical Foundations of Quantum Statistics*, translated by I. Shapiro, Graylock Press, Albany.
Kingman, J.F.C. (1964), The stochastic theory of regenerative events, *Z. Wahrschein-lichkeitstheorie*, **2**, 180–224.
Kolmogorov, A.N. (1949), A local limit theorem for classical Markov chains, *Izvestiya Akad. Nauk S.S.S.R.*, Ser. Mat., **13**, 281–300. *Math. Review*, **11** (1950), 119.
Kumagai, M. (1971), Reliability analysis for systems with repairs, *J. Operations Res. Soc. Jap.*, **14**, 53–71.
Kuznetsov, P.I. and Stratonovich, R.L. (1956), A note on mathematical theory of correlated random points, *Izvestiya Akad. Nauk S.S.S.R.*, Ser. Mat., **20**. Translated by J.A. McFadden in *Selected Translations in Mathematical Statistics and Probability*, **7** (1968), 1–16.
Lampard, D.G. (1968), A stochastic process whose successive intervals between events form a first order Markov chain: I, *J. Appld Prob.*, **5**, 648–68.
Landau, L.D. (1965), *Collected Papers*, Gordon & Breach, New York, pp. 387–91.
Lawrance, A.J. (1970), Selective interaction of a stationary point process and a renewal process, *J. Appld Prob.*, **7**, 483–9.
Lawrance, A.J. (1971), Selective interaction of a Poisson and a renewal process: the dependency structure of the intervals between the responses, *J. Appld Prob.*, **8**, 170–83.
Lecam, L. (1947), Un instrument d'étude des fonctions aléatoires: la fonctionelle caractéristique, *C. R. Acad. Sci. Paris*, **224**, 710–11.
Lévy, P. (1954), Processus semi-Markoviens, *Proc. Int. Congr. Math. Amsterdam*, **3**, 416–26.
Lindley, D.V. (1969), *Introduction to Probability and Statistics from a Bayesian Viewpoint*, Part I, Cambridge University Press.
Loève, M. (1963), *Probability Theory*, 3rd edn, Van Nostrand, Princeton.
McFadden, J.A. (1962), On the lengths of intervals in a stationary point process, *J. R. Statist. Soc.*, **B, 24**, 364–82.
McFadden, J.A. (1965), The mixed Poisson processes, *Sankhyā*, **A, 27**, Part I, 83–92.
McFadden, J.A. and Weissblum, W. (1963), Higher order properties of a stationary point process, *J. R. Statist. Soc.*, **B, 25**, 413–31.
Mercer, A. and Smith, C.S. (1959), A random walk in which the steps occur randomly in time, *Biometrika*, **46**, 30–5.
Messel, H. (1954), The development of a nucleon cascade, in *Progress in Cosmic Ray Physics*, vol. 2, 135–316, North-Holland Publishing Co., Amsterdam.
Moran, P.A.P. (1954), A probability theory of dams and storage systems, *Aust. J. Appld Sci.*, **5**, 116–24.
Moran, P.A.P. (1959), *The Theory of Storage*, Methuen, London.
Moyal, J.E. (1949), Stochastic processes and statistical physics, *J. Roy. Statist. Soc.*, **B, 11**, 150–210.
Moyal, J.E. (1962), The general theory of stochastic population processes, *Acta Math.*, **108**, 1–31.
Nordsieck, A., Lamb, W.E. and Uhlenbeck, G.E. (1940), On the theory of cosmic ray showers: I, *Physica*, **7**, 344–60.
Ornstein, L.S. and Zernike, F. (1914), Accidental derivations of density and opalescence at the critical point of a single substance, *Proc. Kon. Ned. Akad. Sci.* (Amsterdam), **17**, 793–806.

172

Palm, C. (1943), Intensitätschwankungen im Fernsprechverkehr, *Ericsson Technics*, **44**, 1–189.

Parzen, E. (1962), *Stochastic Processes*, Holden-Day (McGraw-Hill, New York).

Prabhu, N.U. (1964), Time dependent results in storage theory, *J. Appld Prob.*, **1**, 1–46.

Prabhu, N.U. (1965), *Queues and Inventories*, John Wiley & Sons, New York.

Prabhu, N.U. (1967), Ladder variables in queueing theory, *J. Math. Phys. Sci.*, **1**, 229–46.

Puri, P.S. (1969), Some new results in the mathematical theory of phage reproduction, *J. Appld Prob.*, **6**, 493–504.

Pyke, R. (1961a), Markov renewal processes: definitions and preliminary properties, *Ann. Math. Statist.*, **32**, 1231–42.

Pyke, R. (1961b), Markov renewal processes with finitely many states, *Ann. Math. Statist.* **32**, 1243–59.

Ramakrishnan, A. (1950), Stochastic processes relating to particles distributed in an infinity of states, *Proc. Camb. Phil. Soc.* **46**, 595–602.

Ramakrishnan, A. (1951), Some simple stochastic processes, *J. Roy. Statist. Soc.*, **B, 13**, 131–40.

Ramakrishnan, A. (1953), Stochastic processes associated with random divisions of a line, *Proc. Camb. Phil. Soc.*, **49**, 473–85.

Ramakrishnan, A. (1954a), A stochastic model of a fluctuating density field, *Astrophys. J.*, **119**, 443–56.

Ramakrishnan, A. (1954b), On the molecular distribution functions of a one-dimensional fluid: I, *Phil. Mag.*, Ser. 7, **45**, 401–10.

Ramakrishnan, A. (1954c), Counters with random dead time, *Phil. Mag.*, Ser. 7, **45**, 1050–2.

Ramakrishnan, A. (1955), Phenomenological interpretation of the integrals of a class of random functions: I and II, *Kon. Ned. Akad. Wet. (Proc. Kon. Ned. Akad. Sci.)*, **58**, 470–82, 634–45.

Ramakrishnan, A. (1958), Probability and Stochastic Processes, in *Handbuch der Physik*, vol. 3, Springer Verlag, Berlin.

Ramakrishnan, A. and Mathews, P.M. (1953), On a stochastic problem relating to counters *Phil. Mag.*, Ser. 7, **44**, 1122–8.

Ramakrishnan, A. and Srinivasan, S.K. (1956), A new approach to cascade theory, *Proc. Ind. Acad. Sci.*, **A, 44**, 263–73.

Reynolds, O. (1895), On the dynamical theory of incompressible viscous fluids and the determination of the criterion, *Phil. Trans.*, **A, 186**, 123–64.

Ricciardi, L.M. and Esposito, F. (1966), On some distribution functions for non-linear switching elements with finite dead time, *Kybernetik*, **3**, 148–50.

Rice, S.O. (1944), Mathematical Analysis of Random Noise: I and II, *Bell Sys. Tech. J.*, **23**, 282–332.

Rice, S.O. (1945), Mathematical Analysis of Random Noise: III and IV, *Bell Sys. Tech. J.*, **25**, 46–156.

Rowland, E.N. (1937), The theory of shot effect, *Proc. Camb. Phil. Soc.*, **33**, 344–58.

Saaty, T.L. (1961), *Elements of Queueing Theory*, McGraw-Hill, New York.

Sankaranarayanan, G. and Swayambulingam, C. (1969), Some renewal theorems containing a sequence of correlated random variables, *Pacific J. Math.*, **30**, 785–803.

Seal, H.L. (1969), *Stochastic Theory of Risk Business*, John Wiley & Sons, New York.

Smith, W.L. (1954), Asymptotic renewal theorems, *Proc. Roy. Soc. Edin.*, **A, 64**, 9–48.

Smith, W.L. (1955), Regenerative stochastic processes, *Proc. Roy. Soc. (Lond.)*, **A, 232**, 6–31.

Smith, W.L. (1957), On renewal theory, counter problems and quasi-Poisson processes, *Proc. Camb. Phil. Soc.*, **53**, 175–93.

Smith, W.L. (1958), Renewal theory and its ramifications, *J. Roy. Statist. Soc.*, **B, 20**, 243–302.

Smith, W.L. (1960), On some general renewal theorems for non-identically distributed variables, *Proc. 4th Berkeley Symp. Math. Statist. and Prob.*, ed. J. Neyman, University of California Press, Berkeley, **2**, 467–514.

173

Srinivasan, S.K. (1955), Applications of Stochastic Processes to Physical Problems, M.Sc. Thesis, University of Madras.

Srinivasan, S.K. (1961), Multiple stochastic point processes, *Zastosowania Matematyki*, **6**, 210–19.

Srinivasan, S.K. (1966), A novel approach to the kinetic theory and hydrodynamic turbulence, *Z. Physik*, **193**, 394–9.

Srinivasan, S.K. (1967), Theory of turbulence, *Z. Physik*, **205**, 221–5.

Srinivasan, S.K. (1968), A Novel Approach to Kinetic Theory of Fluids – Onset of Turbulent Motions, in A. Ramakrishnan (ed.), *Symposia in Theoretical Physics and Mathematics*, **7**, 163–86, Plenum Press, New York.

Srinivasan, S.K. (1969), *Stochastic Theory and Cascade Processes*, American Elsevier Publishing Co., New York.

Srinivasan, S.K. (1971), Stochastic point processes and statistical physics, *J. Math. Phys. Sci.*, **5**, 291–316.

Srinivasan, S.K. (1972), Correlational structure of a pair of interacting point processes, *J. Math. Phys. Sci.*, **6**, 163–79.

Srinivasan, S.K. (1973), Analytic solution of a finite dam governed by a general input. (Submitted to *J. Appld Prob.*)

Srinivasan, S.K. and Gopalan, M.N. (1974), Reliability analysis of an *n*-unit system. (To be published.)

Srinivasan, S.K. and Iyer, K.S.S. (1966), Random processes associated with random points on a line, *Zastosowania Matematyki*, **7**, 221–30.

Srinivasan, S.K. and Koteswara Rao, N.V. (1968), Invariant imbedding technique and age-dependent birth and death processes, *J. Math. Anal. Applic.*, **21**, 43–52.

Srinivasan, S.K. and Rajamannar, G. (1970a), Renewal point processes and neuronal spike trains, *Math. Biosci.* **6**, 331–5.

Srinivasan, S.K. and Rajamannar, G. (1970b), Counter models and dependent renewal point processes related to neuronal firing, *Math. Biosci.*, **7**, 27–39.

Srinivasan, S.K. and Rajamannar, G. (1970c), Selective interaction between two independent stationary recurrent point processes, *J. Appld Prob.*, **7**, 476–82.

Srinivasan, S.K., Rajamannar, G. and Rangan, A. (1971), Stochastic models for neuronal firing, *Kybernetik*, **8**, 188–93.

Srinivasan, S.K. and Ramani, S. (1971), A continuous storage model with alternating random input and output, *J. Hydrology*, **13**, 343–8.

Srinivasan, S.K. and Ramani, S. (1974), A stochastic model of a perpetual inventory *S–s* policy. (To be published.)

Srinivasan, S.K. and Rangan, A. (1970a), Stochastic models for phage reproduction, *Math. Biosci.*, **8**, 295–305.

Srinivasan, S.K. and Rangan, A. (1970b), Age dependent stochastic models for phage reproduction, *J. Appld Prob.*, **9**, 604–16.

Srinivasan, S.K. and Subramanian, R. (1969), Queueing theory and imbedded renewal processes, *J. Math. Phys. Sci.*, **3**, 221–44.

Srinivasan, S.K., Subramanian, R. and Vasudevan, R. (1972), Correlation functions in queueing theory, *J. Appld Prob.*, **9**, 604–16.

Srinivasan, S.K. and Sukavanam, S. (1972), Photo-count statistics of Gaussian light of arbitrary spectral profile, *J. Phys.*, **A**, **5**, 682–94.

Srinivasan, S.K. and Vasudevan, R. (1966), On a class of non-Markovian processes associated with correlated pulse trains and their application to Barkhausen noise, *Nuovo Cimento*, Ser. **A**, **41**, 101–12.

Srinivasan, S.K. and Vasudevan, R. (1967a), Fluctuating density fields, *Ann. Inst. H. Poincaré*, **A**, **7**, 303–18.

Srinivasan, S.K. and Vasudevan, R. (1967b), Fluctuation of photo-electrons and intensity correlation of light beams, *Nuovo Cimento*, Ser. **X**, **47**, 185–93.

Srinivasan, S.K. and Vasudevan, R. (1971), *An Introduction to Random Differential Equations and their Applications*, American Elsevier Publishing Co., New York.

Srinivasan, S.K. and Vasudevan, R. (1972), Photo electron statistics due to mixing of different types of fields, *Nuovo Cimento*, Ser. **B**, **8**, 278–82.

Takacs, L. (1954), Some investigations concerning recurrent stochastic processes of a certain type, *Magyar Tud. Akad. Mat. Kutato Int. Kosl.*, **3**, 115–28.

Takacs, L. (1956a), On secondary stochastic processes generated by recurrent processes, *Act. Math. Hung.*, **7**, 17–29.

Takacs, L. (1956b), On a probability problem arising in the theory of counters, *Proc. Camb. Phil. Soc.*, **52**, 488–98.

Takacs, L. (1956c), On the sequence of events selected by a counter from a recurrent process of events, *Theory of Prob. and Applications* (trans.), **1**, 81–91.

Takacs, L. (1957a), On certain problems concerning the theory of counters, *Act. Math. Hung.*, **8**, 127–38.

Takacs, L. (1957b), On certain sojourn time problems in the theory of stochastic processes, *Act. Math. Hung.*, **8**, 169–91.

Takacs, L. (1967), *Combinatorial Methods in the Theory of Stochastic Processes*, John Wiley & Sons, New York.

Ten Hoopen, M. (1966), Multimodal interval distributions, *Kybernetik*, **3**, 17–24.

Ten Hoopen, M. and Reuver, R.H.A. (1965), Selective interaction of two independent processes, *J. Appld Prob.*, **2**, 286–92.

Ten Hoopen, M. and Reuver, R.H.A. (1967a), On a first passage problem in stochastic storage systems with total release, *J. Appld Prob.*, **4**, 409–12.

Ten Hoopen, M. and Reuver, R.H.A. (1967b), Interaction between two independent recurrent time series, *Information and Control*, **10**, 149–58.

Ten Hoopen, M. and Reuver, R.H.A. (1967c), Analysis of sequences of events with random displacements applied to biological systems, *Math. Biosci.*, **1**, 599–617.

Ten Hoopen, M. and Reuver, R.H.A. (1968), Recurrent point processes with dependent interference with reference to neuronal spike trains, *Math. Biosci.*, **2**, 1–10.

Ter Haar, D. (1952), Gentile's intermediate statistics, *Physica*, **18**, 199–200.

Uhlenbeck, G.E. (1963), Selected Topics in Statistical Mechanics, in Ford, K.W, (ed.), *Statistical Physics – Brandeis Summer Institute* (1962), Benjamin Inc., New York, 1–90.

Uhlenbeck, G.E. (1968), An Outline of Statistical Mechanics, in Cohen (ed.), *Fundamental Problems in Statistical Mechanics*, vol. 2, North-Holland Publishing Co., Amsterdam, 1–29.

Von Mises, R. (1936), Les lois de probabilité pour les fonctions statistiques, *Ann. Inst. H. Poincaré*, **6**, 185–209.

Watson, H.W. and Galton, F. (1874), On the probability of extinction of families, *J. Anthropol. Inst. Great Britain and Ireland*, **4**, 138–44.

Wisniewski, T.K.M. (1972), Bivariate stationary point processes, *Adv. Appld Prob.*, **4**, 296–317.

Wold, H. (1948a), Sur les processus stationnaires ponctuels, *Colloques Internationaux, C.N.R.S.*, **13**, 75–86.

Wold, H. (1948b), Stationary point processes and Markov chains, *Skand. Aktuarietidskr.*, **31**, 229–40.

Yule, G.V. (1924), A mathematical theory of evolution based on the conclusions of Dr J.C. Willis, F.R.S., *Phil. Trans. Roy. Soc. (London)*, **B**, **213**, 21–87.

Yvon, J. (1937), Fluctuations en densité, *Actualités Scientifiques et Industrielles*, Hermann et Cie, Paris.